Raymond W. Henn

Practical Guide to
GROUTING
of Underground Structures

Published by
ASCE Press
American Society of Civil Engineers
345 East 47th Street
New York, New York 10017-2398

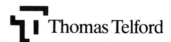

Co-published in the UK by
Thomas Telford Publications
Thomas Telford Services Ltd
1 Heron Quay
London E14 4JD, UK

ABSTRACT

This book presents a hands-on discussion of grouting fundamentals and provides a foundation for the development of practical specifications and field procedures. The author takes a practical approach to the subject of grouting and concentrates on areas such as the types of drilling, mixing and pumping equipment, and their application.The guidebook's primary focus is on the cementitious grouting used in conjunction with the excavation and lining of tunnels, shafts and underground caverns in rock. Overviews of cementitious grouting in soils and chemical grouting are also provided.

Library of Congress Cataloging-in-Publication Data

Henn, Raymond W.
    Practical guide to grouting of underground structures / by Raymond W. Henn
      p.   cm.
    ISBN 0-7844-0140-3
    1. Underground construction. 2. Grouting. I. Title.
TA712.H46 1996   96-3135
624.1'9—dc20   CIP

    The material presented in this publication has been prepared in accordance with generally recognized engineering principles and practices, and is for general information only. This information should not be used without first securing competent advice with respect to its suitability for any general or specific application.
    The contents of this publication are not intended to be and should not be construed to be a standard of the American Society of Civil Engineers (ASCE) and are not intended for use as a reference in purchase specifications, contracts, regulations, statutes, or any other legal document.
    No reference made in this publication to any specific method, product, process or service constitutes or implies an endorsement, recommendation, or warranty thereof by ASCE.
    ASCE makes no representation or warranty of any kind, whether express or implied, concerning the accuracy, completeness, suitability or utility of any information, apparatus, product, or process discussed in this publication, and assumes no liability therefore.
    Anyone utilizing this information assumes all liability arising from such use, including but not limited to infringement of any patent or patents.

Photocopies. Authorization to photocopy material for internal or personal use under circumstances not falling within the fair use provisions of the Copyright Act is granted by ASCE to libraries and other users registered with the Copyright Clearance Center (CCC) Transactional Reporting Service, provided that the base fee of $4.00 per article plus $.25 per page is paid directly to CCC, 222 Rosewood Drive, Danvers, MA 01923. The identification for ASCE Books is 0-7844-0140-3/96 $4.00 + $.25 per page. Requests for special permission or bulk copying should be addressed to Permissions & Copyright Dept., ASCE.

Copyright © 1996 by the American Society of Civil Engineers,
All Rights Reserved.
Library of Congress Catalog Card No: 96-3135
ISBN 0-7844-0140-3
Manufactured in the United States of America.

Co-published in the UK by Thomas Telford Publications, Thomas Telford Services Ltd, 1 Heron Quay, London E14 4JD, UK.

Cover photo by George Sarek.

CONTENTS

Preface vii

1. Introduction 1
    Purpose 3
    Scope 3

2. Grouting Methods 5
    2.1 Grouting in Soil 6
        2.1.1 Jet Grouting 6
        2.1.2 Compaction Grouting 11
        2.1.3 Permeation Grouting 17
        2.1.4 Hydrofracture Grouting 21
    2.2 Grouting in Rock 22
        2.2.1 Consolidation Grouting 23
        2.2.2 Curtain Grouting 23
    2.3 Structural Grouting 24
        2.3.1 Contact Grouting 24
            2.3.1.1 Cast-in-Place Concrete Lining 26
            2.3.1.2 Mapping Cracks in Concrete Lining 30
            2.3.1.3 Precast Concrete Lining 32
            2.3.1.4 Steel and Cast Iron Segment Linings 36
            2.3.1.5 Steel Penstock Lining 38
            2.3.1.6 Tunnel and Station Linings Using
                      Waterproofing Systems 39
        2.3.2 Embedment Grouting 43
        2.3.3 Prestressing 44
        2.3.4 Cellular Concrete 49
    2.4 Shaft Grouting 52
        2.4.1 Pregrouting of Shaft from Ground Surface 53
        2.4.2 In-Shaft Grouting 56
    2.5 Probing and Grouting ahead of Tunnel and
        Chamber Excavations 60
        2.5.1 Probe Holes 60
        2.5.2 Grouting ahead of Excavations 61

**3. Geotechnical Considerations in Grouting
Program Planning** 65
    3.1 Basic Considerations 66

    3.2 Geotechnical Investigation and Geotechnical
         Design Summary 66
    3.3 Grouting to Limit Groundwater Infiltration 70
    3.4 Grouting to Increase Stability 75
    3.5 Grouting to Increase Strength 80
    3.6 Groutability Ratio 80
    3.7 Other Geotechnical Considerations 81

4. Equipment 83

    4.1 Drills 84
        4.1.1 Percussion Drilling 86
        4.1.2 Rotary Drilling 90
        4.1.3 Down-Hole-Drill (Down-Hole-Hammer) Drilling . 92
    4.2 Mixers 93
    4.3 Agitators 98
    4.4 Water Meters 100
    4.5 Pumps 101
    4.6 Pressure Gauges 103
    4.7 Gauge Savers 103
    4.8 Packers 104
    4.9 Nipples 107
    4.10 Delivery and Distribution System 109
    4.11 Data Acquisition and Recording Equipment 110
    4.12 Automated Batching Systems 113
    4.13 Grouting Jumbos 117
    4.14 Equipment Configuration 117

5. Grouting Materials 121

    5.1 Portland Cement 121
    5.2 Ultrafine Cement 123
    5.3 Sand 124
    5.4 Admixtures 124
        5.4.1 Dispersants 126
        5.4.2 Accelerators 126
        5.4.3 Gas-Producing Agents 127
    5.5 Water 127
    5.6 Bentonite 128

6. Specifications 129

    6.1 Applicable ASTM's and Other Standards 130
    6.2 Developing Grouting Specifications 130

7. Grout Hole Layout ... 133

7.1 Grout Hole and Grout Ring Spacing ... 133
    7.1.1 Contact Grouting ... 133
    7.1.2 Consolidation Grouting ... 140
    7.1.3 Curtain Grouting ... 142
7.2 Hole Depth and Diameter ... 145

8. Grout Placement Operations ... 147

8.1 Proportioning ... 147
    8.1.1 Water:Cement Ratios ... 147
    8.1.2 Sand ... 148
    8.1.3 Admixtures ... 149
    8.1.4 Bentonite ... 149
8.2 Delivery Pressure ... 149
8.3 Refusal Criteria ... 155
8.4 Crew Size and Organization ... 155
8.5 Production ... 155
8.6 Safety and Environmental Issues ... 157
    8.6.1 Personnel Safety ... 157
    8.6.2 Waste Disposal ... 159

9. Field Quality Control ... 161

9.1 Preconstruction Checklists ... 162
9.2 Certifications and Test Reports ... 162
9.3 Inspection Reports ... 162
    9.3.1 Drilling Reports ... 164
    9.3.2 Grouting Reports ... 164
9.4 Testing ... 164
    9.4.1 Laboratory Testing ... 164
    9.4.2 Field Testing ... 167

10. Chemical Grouting ... 169

10.1 Applications ... 169
10.2 Types of Chemical Grouts ... 170
10.3 Equipment ... 171
    10.3.1 Batching Equipment ... 171
    10.3.2 Injection Equipment ... 172
10.4 Design of a Chemical Grouting Program ... 174
    10.4.1 Groutability of the Ground ... 174

          10.4.2 Grouting Program Design 175
     10.5 Chemical Grouting Case Histories 176
          10.5.1 Chemical Grouting from the Surface 176
          10.5.2 Chemical Grouting from the Tunnel
                  Working Face 179

References **183**

Index **187**

PREFACE

Grouting is a part of most underground civil engineering and mining projects. Depending on the type and operating parameters of the underground facility, geology, and groundwater conditions, grouting can represent a sizable cost and scheduling component of a project. This book is written as a practical guide for engineers, construction supervisors, inspectors, and others involved in the planning, design, and implementation of underground grouting programs. Its primary purpose is to present a hands-on discussion of grouting fundamentals and to provide a foundation for the development of practical specifications and field procedures for underground grouting applications. The reader is encouraged to combine the information in this book with other information sources on grouting. An excellent companion book to this one is *Dam Foundation Grouting*, by Ken Weaver, published in 1991 by the American Society of Civil Engineers.

Cementitious grouting used in conjunction with the excavation and lining of tunnels, shafts, and underground caverns in rock is the primary focus of this work. Overviews of cementitious grouting in soils and chemical grouting are also provided.

The guide takes a practical approach to the subject of underground grouting by concentrating on areas such as grouting methods; types of drilling, mixing, and pumping equipment; and their application. Grouting materials and specifications, record keeping, quality control and testing requirements, field operations, and production rates are also covered. These are all important elements to the overall success of a grouting program, yet they are usually not addressed in detail in engineering literature on grouting.

The guide is intended to compliment existing engineering literature by presenting grouting equipment, technology, and methodologies that are presently available. This information will enhance the planning, field implementation, and quality of an underground grouting program. A better understanding of the capabilities and limitations of the art and science of grouting can only improve the overall results and performance of a grouting program. The text is written assuming that the reader has a basic knowledge of geology, engineering principles, and underground construction methods.

Chapter 3, "Geotechnical Considerations in Grouting Program Planning," was contributed by Thomas M. Saczynski. Chapter 10, "Chemical Grouting," was contributed by Daniel F. McMaster and Michael J. Robison. The computer-generated graphics were created by Vladimir Goubanov. Much of the photographic work was done by Thomas Jenkins. The author would like to thank Pamela Moran, Patricia Henn, and Joe Sperry for their review of the draft manuscript.

CHAPTER 1

INTRODUCTION

The development of cementitious permeation grouting got its start as a method to improve the foundation material of civil engineering structures built in and around bodies of water. The concept of injecting self-hardening cementitious slurry was first employed in 1802 in France to improve the bearing capacity under a sluice (Bruce 1995). The development of cement grouting continued in France and England throughout the 1800s. The applications were concentrated on civil structures such as canals, locks, docks, and bridges (Bruce 1995).

The first recorded use of cementitious grout in underground construction was when, in 1864, Peter Barlow patented a cylindrical one-piece tunnel shield with a cast iron liner constructed from within. The annular void left by the tail of the shield was filled with grout (Tirolo 1994). In 1893 the first systematic grouting of rock in the United States was performed at the New Croton Dam, in New York (Weaver 1991). The grouting program at the Hoover Dam between 1932 and 1935 is said to mark the beginning of systematic design of grouting programs in the United States (Glossop 1961).

The development and advances of underground grouting technology in soil and rock as they apply to design, equipment, and materials have for the most part paralleled the advances made in dam and foundation grouting preformed from the surface. Today, most underground civil engineering and mining projects require some form of grouting. Depending on the type and operating parameters of the underground facility, the geology, and groundwater conditions, a grouting program can represent a considerable cost and scheduling component of a project.

Grouting performed in conjunction with engineered underground structures, such as tunnels, shafts, chambers, and mine workings, is similar to grouting operations performed from the surface, such as installing a grout curtain for a dam. In both cases grouting is used to fill pores, fissures, or voids in the host geologic materials to reduce seepage, to strengthen foundation material, or to improve ground-structure interaction.

While grouting performed from the surface and underground grouting share many similarities, limitations on access and work space restrictions are particularly significant aspects in underground grouting. Logistical considerations of the underground grouting operation must include the following:

- Choosing a method to transport the grout materials and equipment to the work area and scheduling delivery
- Designing the grouting program as it pertains to hole depth, spacing, and orientation to best fit the underground opening geometry and geologic conditions
- Selecting the proper size and layout of equipment and sizing of work crews based on the limitations presented by access and the physical space available in the work area
- Developing methods to remove grout waste and wash water from underground without clogging the dewatering pumps and piping systems
- Minimizing disruptions to other underground construction and mine production activities

While these issues may appear to be more of a planning and coordination effort, they are likely to affect adversely the quality, cost, and schedule of the grouting program if they are not properly addressed before grouting operations begin. Additionally, a poorly planned and executed grouting program can cause increased costs and schedule delays in other activities of the project. For example, it is common on hydroelectric and water resource projects for the underground grouting program to be on the critical path of the project schedule. Likewise, grouting performed to control inflows of groundwater during shaft sinking for a mine development project is normally a critical-path activity.

A grouting program is usually a pre-engineered part of the final design of a project. Methods specifications, as opposed to performance specifications, are most commonly used for underground grouting operations. The contract specifications dictate the grouting methods, grout materials, mix designs, hole size, injection pressure, and refusal criteria to be used. The contract drawings show the grout hole locations, geometry, spacings, and depths.

Additionally, contractor designed grouting programs are commonly used by excavation and mine development contractors to improve the construction process. For example, a contractor may elect to grout prior to or during excavation in order to stop or substantially reduce the flow of water into the excavation. This reduces dewatering costs and increases construction efficiency. Grouting may also be used to modify the behavior of the material to be excavated. Solidifying loose or running soil and stabilizing blocky rock conditions prior to excavation are good examples.

INTRODUCTION

Cement-based grouts, also referred to as "cementitious" or "slurry" grouts, are the most widely used in underground applications in rock. Cement-based grouting in rock is the primary focus of this guide. Overviews of cementitious grouting in soil and chemical grouting in soil and rock are also provided.

PURPOSE

This book is written as a practical guide for engineers, construction supervisors, inspectors, and others involved in the planning, design, and implementation of underground grouting programs. A need for a pragmatic, hands-on reference has been recognized by practitioners who perform as contractors, engineers, and consultants engaged in underground works. The guide is thus intended to bridge the gap between the existing technical literature and the applications as presently practiced in the field. The information herein should be used in conjunction with other sources on grouting. An excellent companion book is *Dam Foundation Grouting*, by Ken Weaver, published in 1991 by the American Society of Civil Engineers.

SCOPE

This guide is presented in 10 chapters, the first of which serves as an introduction. The second chapter gives an overview of the types of grouting methods with respect to the geological medium in terms of soil grouting versus rock grouting. Subsequent chapters deal with the geotechnical considerations in planning a grouting operation; an overview of the types, sizes, and functions of grouting equipment; descriptions of typical grouting materials; a review of applicable standards and codes and an example of a grouting specification; a discussion of drilling including hole diameter, depth, and spacing; a treatment of field placement operations that focuses on proportioning, mixing, delivery pressures and refusal criteria, crew size, and safety and environmental issues; quality control; and an overview of chemical grout.

CHAPTER 2

GROUTING METHODS

The first step in selecting a grouting method is to determine the general category of the geologic material to be grouted: it is either soil or rock. It is quite simple to envision a material consisting of either soil or rock and intuitively realize the distinction between the two. A clear definition, however, is necessary. The contract documents must give the definitions of the various earth materials expected to be encountered on the project as well as their expected behaviors, keeping in mind that existing technical literature may contain some overlap of the definitions. Variations in definitions also exist among different agencies and from one geographical region to another. There may also exist gray areas such as when the material could be either a weak rock or a strong clay, or weathered rock versus a true soil.

For the purpose of this guide, rock is defined as the hard and solid formations of the earth's crust (Jumikis 1983). Soil is defined as sediments or other unconsolidated accumulations of solid particles produced by the physical and chemical disintegration of rocks, which may or may not contain organic matter (The Asphalt Institute 1978).

Although the simple definitions for soil and rock are adequate, choosing a grouting method is more complex and requires additional consideration. The desired function of the grout, for example, stabilizing ground or cutting off water, is an indispensable aspect. The character of the earth material to be grouted, with respect to the size and spacing of the discontinuities in rock or the particle size of the soil, is also critical.

A single underground project may require several different grouting methods. These various methods may require different equipment and may be performed at different stages and time periods in the project. For example, stabilization of a tunnel portal in soil may require jet grouting prior to the start of the actual tunnel excavation. Later, the cast-in-place concrete tunnel lining may require contact grouting to fill the space, or void, left between the concrete and rock after the liner concrete has been placed. A period of months, or even years, may transpire between the application of these two different grouting operations. They may also be performed by two different contractors.

Additionally, a specific operation, such as tunnel excavation, often requires excavating in soil (soft-ground) and in rock (hard-rock) at different locations along the same tunnel alignment. In some cases both soil and rock are encountered simultaneously within an excavation. An example of this is when soil is present in the upper portion of the working face of a tunnel excavation and rock is encountered in the lower portion. This condition is called "mixed-face" tunneling. In such cases, since the upper portion of the tunnel is in soil, different grouting methods for the upper and lower portions may be required. Even when a project is totally founded in either soil or rock, the geologic material itself is likely to exhibit changes along the alignment in soil gradation, rock type, structure, and groundwater conditions. These varying conditions may also necessitate use of different grouting methods.

This chapter discusses common grouting methods used in underground applications in both soil and rock. Jet, compaction, permeation, and hydrofracture grouting methods are frequently used in soft-ground excavations, whereas hard-rock excavations regularly require consolidation grouting methods. Contact grouting is common in both soft-ground and rock excavation when a lining such as concrete or steel is installed over the excavated material. Curtain and embedment grouting are less routine and are used primarily in conjunction with the construction of hydraulic structures such as hydroelectric developments and water conveyance systems. Prestress grouting is a specialized application used primarily in high-pressure water conveyance shaft and tunnel construction.

2.1 GROUTING IN SOIL

Grouting of soil in conjunction with underground structure can be performed from the surface or from underground, usually at the working face. Factors such as soil gradation, groundwater levels, depth of the structure below the surface, and surface access for grouting equipment will determine which approach or combination of approaches is best for a specific project. Chemical grouting of soil is discussed in Chapter 10.

2.1.1 Jet Grouting

Jet grouting is a relatively new technique developed since the 1960s (Welsh 1991) in which thin high-pressure jets of cement grout are discharged laterally into a borehole wall. The jet simultaneously excavates and mixes with the soil. The cutting action of the jets can be enhanced by incorporating compressed air. The high-pressure injection of the grout slurry through the nozzle allows changes in the geotechnical characteristics of the soil to suit the specific requirements of the project. The result is a column of modified soil with low permeability and improved strength.

One type of jet grouting method proceeds in two phases, the drilling phase and the withdrawal phase with simultaneous programmed injection (Fig. 2-1).

GROUTING METHODS

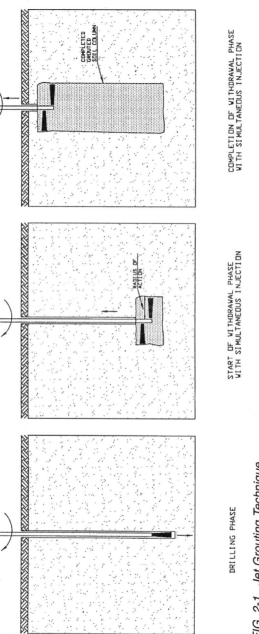

FIG. 2-1. Jet Grouting Technique

The radius of the jet-grouted column, called the radius of action, depends on several factors, such as

- Work pressure, which is generated by a special pump that can provide from 1 to 800 bars (14.7 to 11,760 psi) of pressure
- Injection time, which is determined by the rate at which the drilling rods are rotated and withdrawn
- Shear strength of the in situ soil
- Size of the nozzles
- Specific weight of the grout slurry

There are many possible applications for using jet grouting in underground construction. Some of the most common uses are shown in Figs. 2-2, 2-3, and 2-4.

For Contract E, on the Islais Creek Transport/Storage Project, in southeast San Francisco, jet grouting was used extensively to modify the in situ soil conditions along the tunnel alignment (Burke 1995). The project required the excavation and lining of two relatively short soft-ground tunnels. A 4.1 m (13.5 ft) diameter tunnel of approximately 76 m (250 ft) in length and a 4.6 m (15 ft) diameter tunnel of approximately 155 m (510 ft) in length. The soil conditions along the alignment were predominantly Bay Mud, which is a plastic silty clay or clayey silt with trace amounts of organic shells and sand. It is generally classified as CH or MH according to the Unified Soil Classification System.

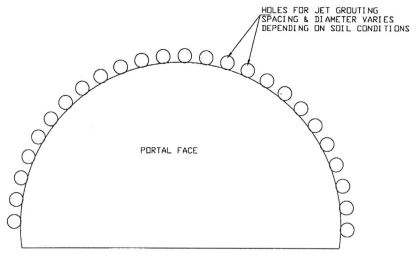

FIG. 2-2. Jet Grouting Used for Portal Development in Soft-Ground Tunneling

GROUTING METHODS

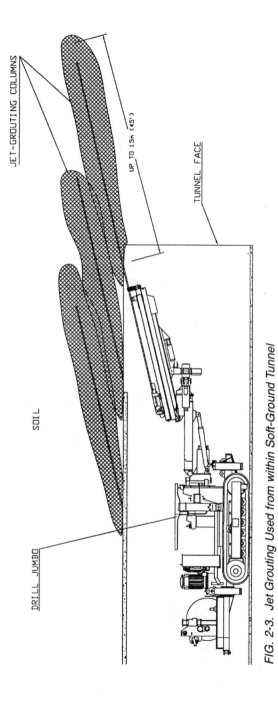

FIG. 2-3. Jet Grouting Used from within Soft-Ground Tunnel

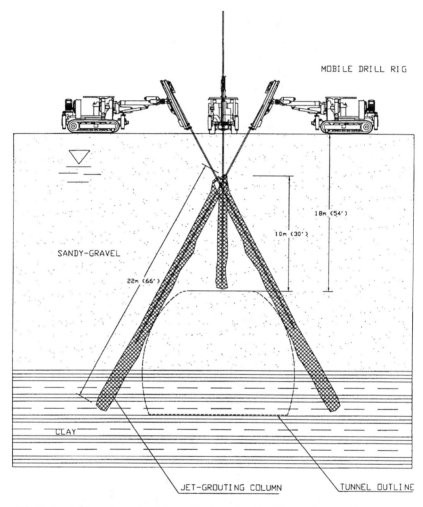

FIG. 2-4. *Jet Grouting Used from Surface for Soft-Ground Tunneling*

The project specifications called for compressed air tunneling to resist water inflows and to reduce squeezing ground and other ground movements. The primary ground support was to be liner plates erected within the shield with contact grouting behind the plates. The contractor proposed the elimination of compressed air completely in favor of soil modification by jet grouting. The contractor's proposal was accepted. The jet grouting consisted of overlapping 2 m (6 ft) diameter grout columns having compressive strength ranging from 4.1 to 6.9 MPa (600 to 1,000 psi). In addition to encompassing the entire tunnel cross section, the jet-grouted columns extended approximately 2 m (6 ft) outside the circumference of the tunnel.

The change from compressed air to jet grouting required a change in methods of tunneling and equipment. To excavate the modified soil the contractor used a road header boom mounted within the shield. Primary support was steel ribs and lagging. The final lining was a 3.2 m (10.5 ft) inside diameter (ID) steel pipe for the 4.1 m (13.5 ft) tunnel and a 3.7 m (12 ft) ID steel pipe for the 4.6 m (15 ft) tunnel. Cellular concrete was used as backfill between the pipe and steel ribs and lagging in both cases. The project was completed to everyone's satisfaction (Burke 1995).

Jet grouting is a specialized form of grouting requiring special equipment and procedures. It is therefore performed by a limited number of contracting firms. These specialty contractors usually design and construct the jet-grouting system to meet the design engineer's or the prime contractor's requirements. Specialty contracting firms should be contacted early in the design process to maximize use of the method. A performance specification, as opposed to a method specification, is also advised when jet grouting is specified.

2.1.2 Compaction Grouting

Compaction grouting is a technique of injecting a stiff grout with a 25–50 mm (1–2 in.) slump through pipes or casings that are driven into soil. The injection pipe diameter should be kept between 50 and 100 mm (2 and 4 in.). Typically, hole spacing ranges from 1.5 to 6.1 m (5 to 20 ft) on centers depending on the soil conditions of the project and desired results of the grouting. The grout is pumped at relatively high pressures between 1 and 7 MPa (150 and 1,000 psi) (Baker 1985). The grout exiting the bottom of the pipe forms a bulb-shaped mass that increases in volume, thereby densifying the soft, loose, or disturbed soil surrounding the mass (Fig. 2-5).

Because of the low slump, the grout does not enter the pores of the soil but remains in a homogeneous mass, thus allowing the controlled displacement of grout to compact the loose soil surrounding it. The influence of the compaction grout mass can extend well beyond the grout mass itself, involving 20 times the grout volume (Baker 1985). A typical compaction grout mix might consist of 10% portland cement, 25–30% silt and sand, and sufficient water to provide the proper slump. Design mixes may also contain such ingredients as fly ash, gravel, bentonite, and water-reducing agents.

Compaction grout holes can be drilled vertically or, where existing structures or other restrictions preclude this, they can be angled (Figs. 2-6 and 2-7). Holes are drilled in accordance with a predetermined pattern to a required depth.

When performing compaction grouting adjacent to an existing structure or one under construction, it is necessary to carefully monitor the structure

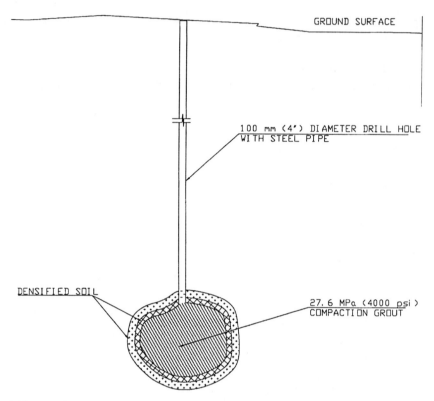

FIG. 2-5. Typical Compaction Grouting Showing Grout Bulb and Densified Soil around Bulb

for displacement or increased loads imposed by the grouting operation. Tunnel steel sets can be buckled, and lagging boards can snap from the increased loads. Also, tunnel-boring machines can be distorted into an egg shape and immobilized as a result of compaction grouting.

The Bolton Hill Tunnel, constructed between 1977 and 1980, is reported to be the first application of compaction grouting for controlling ground movement during tunneling (Baker et al. 1983). The 3.5 km (11,441 ft) long, 5.9 m (19.3 ft) diameter, twin steel-lined tunnels are part of the Northwest Line of the Baltimore Regional Rapid Transit System. They are located between 12.2 and 22.9 m (40 and 75 ft) below the surface in very dense sand and gravel, with occasional silt and clay lenses, overlying very dense residual soil.

During the project, compaction grouting was specified by contract to restore 40 structures to their original position. The buildings were principally two- and five-story-high structures with brick bearing walls. The contractor

GROUTING METHODS

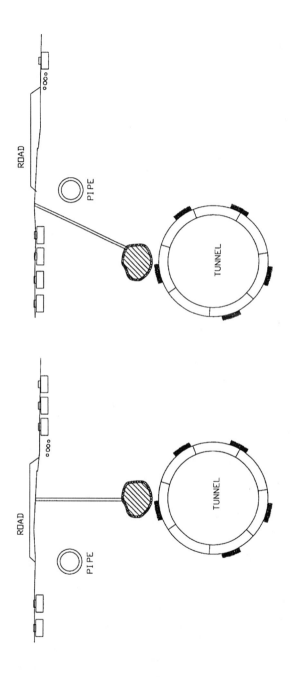

FIG. 2-6. *Section View of Compaction Grouting with Holes Drilled (A) Vertically; and (B) Angled*

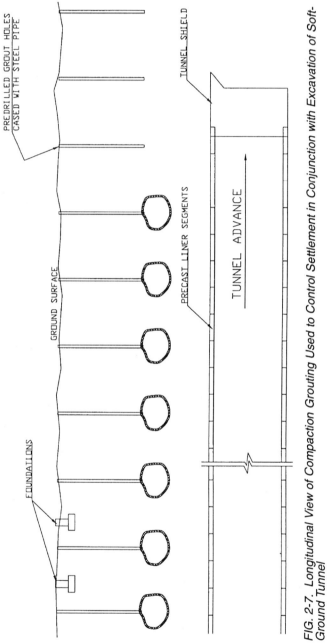

FIG. 2-7. Longitudinal View of Compaction Grouting Used to Control Settlement in Conjunction with Excavation of Soft-Ground Tunnel

selected a 5.9 m (19.3 ft) diameter articulated tunnel shield with an excavator arm. The shield was equipped with breasting shelves and a large muck pan to support the soil in the open face of the shield. Compressed air at 0.4–0.6 bars (6–9 psi) was used to control sand and water inflows at the face.

The original specifications called for compaction grout pipes to be pre-installed in the vicinity of individual buildings. The compaction grouting would begin when building settlement exceeded 6.3 mm (0.25 in.). This scheme would have required placement of many grout pipes from within building basements. It also would have entailed an extensive settlement monitoring schedule with grout crews on continuous standby.

The grouting subcontractor to the tunnel contractor proposed an alternate procedure. The proposal, which was accepted, was to install grout pipes from the edge of the street surface and place compaction grout bulbs approximately 1.3–4.5 m (5–15 ft) above the tunnel crown. This would limit movement closer to their source before they could affect the surface structures.

Grouting was started on a hole as the tail shield of a machine passed 1.5–2.1 m (5–7 ft) beyond the preinstalled vertical and inclined 75 mm (3 in.) grout pipes. The grout had a slump of 25–50 mm (1–2 in.) and was injected at pressures between 24.8 and 41.4 bars (360 and 600 psi). The first 92 m (300 ft) of the tunnel alignment was instrumented. The first tunnel heading was advanced, then stopped. The second heading was then driven to evaluate settlement under both single- and double-heading conditions. Grout holes were first spaced 1.5 m (5 ft) on centers. This spacing was increased to 3 m (10 ft) based on test results. Recorded surface settlements were 9.2 mm (0.4 in.) above the first tunnel and 12.6 mm (0.5 in.) above the second tunnel.

Another tunnel, the Papago Freeway Drainage Tunnel in Phoenix, used both compaction and chemical grouting during tunneling to control surface settlement, ground runs, and the development of large, open chimneys to the ground surface.

The project was constructed between 1984 and 1987. It consisted of three soft-ground tunnels. These tunnels were referred to as the north, east, and west tunnels. The total length of the tunnels was approximately 10.5 km (6.5 m). Of the three tunnels, only the west tunnel was grouted. It was 4.3 km (13,970 ft) long with an excavated diameter of 7.6 m (25 ft) and a finished diameter of 6.4 m (21 ft). The crown of the tunnel was between 10.1 and 13.7 m (33 and 45 ft) below the surface. All three tunnels were excavated through coarse alluvial sand-gravel-cobble deposits in downtown Phoenix. All but a third of the east and west tunnel lengths were driven above the groundwater table. In the portion of tunnel excavation below the groundwater table, the groundwater levels were drawn down to below tunnel invert using several large-diameter, high-capacity wells.

The contractor selected a shield machine with a digger arm. Almost immediately after starting the east tunnel, which had no ground modification associated with the excavation, a series of large ground runs occurred. The runs resulted in large surface settlements on the order of 2–3 m (6–10 ft) and in large, open chimneys to the ground surface. Since the tunnel alignment ran under highway right-of-way with little surface development, no surface structural damage was sustained. In an effort to control these runs, the contractor modified the tunnel machine by adding poling plates and breasting boards. The east tunnel was completed with these modifications.

Based on the ground movements experienced during the east tunnel drive, and the fact that the west tunnel alignment passed under highly developed properties, a ground modification program was considered. A risk assessment was performed based on the types of development along the west alignment. The development included streets and pavements, high-voltage electric lines, high-pressure gas lines, and fiber-optic telecommunications cable, as well as residential and commercial buildings. Based on the risk assessment and the experience gained on the east tunnel drive, a program of compaction and chemical grouting was developed for the west tunnel.

The compaction grouting was performed from the surface as the tunnel advanced. Grout holes were drilled on 3 m (10 ft) centers along the centerline of the tunnel alignment to within approximately 3 m (10 ft) above the crown of the tunnel. Each hole was fitted with a 75 mm (3 in.) diameter grout pipe. Compaction grouting was started on a hole just after the tail shield passed beyond a grout pipe. At first, noncementitious grout was used consisting of well-graded silty sand, fly ash, and water. The mixture had a slump of approximately 50 mm (2 in.). The mix was later changed to a readily available native sandy silt without fly ash. This mix was used for the remainder of normal tunneling operations. However, when ground losses in excess of 76.5 m^3 (100 cu yd) occurred at the face, one bag of portland cement was added to every 1.5 m^3 (2 cu yd) of grout mixed. The cement was added to strengthen the ground and reduce the chances of grout flow into the heading around the face of the shield. Lyman (1988) reported the grouting program was effective in limiting near-surface ground movement. This effectively minimized the risk of damage to utilities, street pavement, and nearby buildings throughout most of the west tunnel alignment.

As with jet grouting, compaction grouting is a specialized type of grouting requiring special equipment and procedures. It is therefore performed by a limited number of contracting firms. These specialty contractors usually design and construct the compaction grouting system to meet the design engineer's or the prime contractor's requirements. As with jet grouting, to maximize use of this method specialty contracting firms need to be contacted early in the design process. A performance specification, as opposed to a method specification, is advised when compaction grouting is specified.

2.1.3 Permeation Grouting

In permeation grouting, grout is injected into the pore spaces of the soil. The technique is used to control water or to improve the structure of the soil. It is the oldest and remains the most common type of grouting used.

A typical cementitious grout mix consists of cement, bentonite, and water. Fillers such as fly ash and fine sand are often used when grouting gravel or coarse sandy gravel to reduce material costs. Types I and II portland cement are most commonly used. Type III high-early portland cement and ultrafine cements are used to grout fine-graded soil. Fig. 2-8 shows gradation curves for conventional, high-early, and ultrafine cements.

The size and distribution of the in situ soil particles determine the groutability of the foundation material, but they vary widely (Fig. 2-9). The designer or contractor can, however, select the type grout with appropriate consideration given to the particle size of the cement and its relation to the soil to be grouted. As the in situ soil gradation becomes finer, cement-based grouts diminish in their effectiveness to penetrate the soil mass. Beyond a certain minimum soil particle size, portland cement grouts are totally ineffective (Fig. 2-10), and ultrafine cements and chemical grouts are used. In some cases, even ultrafine and chemical grouts do not penetrate the soil mass when the particle size exceeds a certain fineness. Chapter 3 discusses the concept of a "groutability ratio," which provides a relationship between the soil and grout particles. Chapter 10 discusses chemical grouting in more detail.

An example of an application of permeation grouting is the installation of the so-called barrel vault method of soil stabilization, used during excavation of the Schurzeberg Tunnel in Oberrieden, Germany (Deinhard et al. 1991). As part of the Schurzeberg project, a 100 m² (1,100 ft²) by 68 m

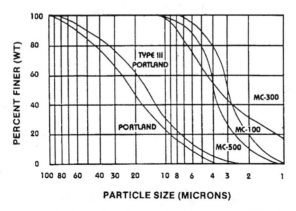

FIG. 2-8. Gradation Curves for Conventional, High-Early, and Ultrafine Cements (Clarke 1995)

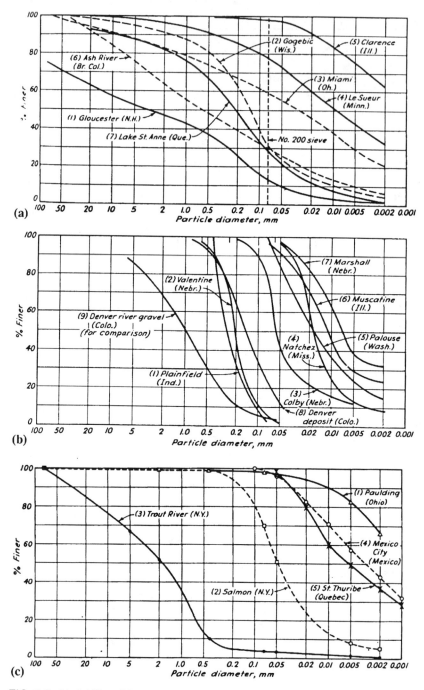

FIG. 2-9. Variability of Gradations for Different Soils (Foundation Engineering, Second Edition, by R.B. Peck, W.E. Hanson, and T.H. Thornburn. Copyright ©1974. Reprinted by permission of John Wiley & Sons, Inc.)

GROUTING METHODS

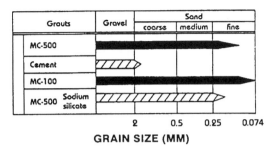

FIG. 2-10. Groutability Based on Grout Type versus Soil Particle Size (Clarke 1995)

(225 ft) long two-lane road tunnel had to be excavated through a 120-year-old man-made embankment. The embankment carries two railway tracks. When the embankment was constructed, whatever materials that were on hand were used, with little regard for adequate compaction.

The tunnel was excavated using the New Austrian Tunneling Method (NATM). The NATM design called for the tunnel crown, or roof material, to be stabilized using multiple rows of pre-excavation permeation grouting (Fig. 2-11). The grouting was performed from the two portals, one on each

FIG. 2-11. Portal Development Showing Three Rows of Permeation Grout Holes (Courtesy of Dr. G. Sauers Corp. 1995)

side of the embankment, with the grouted zones overlapping in the center of the embankment. Each of the portals had a total of 82 grout holes with manchette tubes installed in a three-row pattern. A typical manchette tube assembly is shown in Figs. 2-12 and 2-13. In the first row the tubes were installed at 600 mm (24 in.) centers about 300 mm (12 in.) outside of the excavation limits. In the second row, 500 mm (20 in.) outside the first, they were installed at 1.2 m (4 ft) centers. The third row was approximately 1 m (3 ft) outside the second row, with tube spacing of 1.5 m (5 ft). The cover between the crown of the tunnel and the top of the rail track was 10 m (33 ft) maximum.

The manchette tube installation consisted of drilling a 133 mm (5-15/16 in.) outside diameter (OD) by 35 m (115 ft) long cased hole. A 74 mm (2-5/16 in) OD with a 67 mm (2-5/8 in.) ID manchette tube having one sleeve port per linear meter (3 ft) was used. A bentonite slurry was injected into the annular space between the tube and the casing during the casing removal. The bentonite slurry was then displaced by pumping a weak cement grout. Next, using a double packer arrangement, permeation grout was injected through the manchette tube. A starting pressure of 20 bars (294 psi) was used to break through the rubber sleeve covering openings in the manchette tube and crack through the weak cement-grout annulus. Once

FIG. 2-12. Manchette Tube Assembly Showing Multiple Grout Port Sleeves

GROUTING METHODS

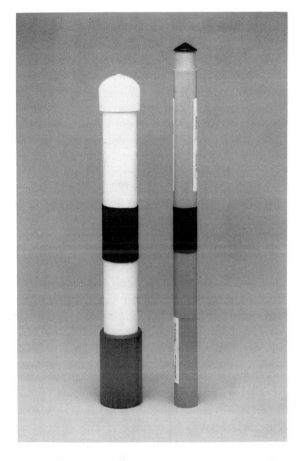

FIG. 2-13. Manchette Tube Assembly Showing Close-Up of Grout Port Sleeve

grout began to flow into the surrounding embankment material, the injection pressure was reduced to 2–5 bars (29.4–73.5 psi).

2.1.4 Hydrofracture Grouting

The hydrofracture, or soil-fracture, grouting method involves locally confined and controlled fracturing of a soil unit by injecting a stable but fluid, cement-based grout at high pressure, for example, up to 4 MPa (580 psi) (Bruce 1995). The method is used primarily to increase the bearing capacity and shear resistance of the soil. It can be used before tunnel excavation begins to raise structures and compensate for anticipated settlement. It can also be used contemporaneously with soft-ground tunneling to offset the settlement (Drooff et al. 1995).

Until recently, the use of soil fracturing had been limited to the French grouting industry. The first successful application of soil fracture grouting in North America was on the St. Clair River Tunnel (Drooff et al. 1995). On the St. Clair project a soil-fracturing method trade-named "Soilfrac" was used. The project consisted of excavating a 9.2 m (30 ft) diameter soft-ground tunnel using an earth pressure balance (EPB) tunneling machine. The tunnel alignment passed under an oil refinery research building approximately 7 m (23 ft) below its footings. The predicted settlements in the soft St. Clair till were in excess of 100 mm (4 in.). The project specifications required maintaining the research building within 10 mm (3/8 in.) of its original position.

The soil-fracturing program consisted of two phases. In the first phase, called "precondition grouting," soil-fracture grouting was used to reinforce the soil below the research building and preheave or raise the building 5 mm (3/16 in.) to compensate for the anticipated settlement. The phase 1 holes were drilled horizontally under the research building from two work shafts installed specifically for the purpose. Two layers of fracture grouting holes were drilled from each shaft. A total of four layers of grout holes overlapped under the research building. The grout pumps for the project were developed to allow for a controlled injection of cement grout with a flow rate of between 4 and 30 L/min (0.14 and 1.05 cfm) with a maximum operating pressure of 70 bars (1029 psi) (Droof et al. 1995).

Phase 2 of the program was to use fracture grouting during the actual tunneling operation beneath the research building to compensate for settlement. The building was heavily instrumented with electrolevels to measure and record settlement during tunneling. The instruments recorded a slight heave of 2–3 mm (1/16 to 1/8 in.) as the EPB machine mined underneath the research building. This heave disappeared when the mining operation passed beyond the building. Subsequently, after 24 hours, some settlement did occur. The settlement was on the order of 4–6 mm (3/32 to 1/4 in.). However, the research building still had a net heave of 4–6 mm (3/32 to 1/4 in.). Because of these favorable results it was not necessary to perform the phase 2 fracture grouting underneath the research building.

2.2 GROUTING IN ROCK

Grouting of rock in conjunction with underground structures can be performed from the surface or from underground. Factors such as the geological character of the rock, depth of structure, groundwater levels, and surface access for grouting equipment will determine which approach or combinations of approaches is best for a specific project. Most grouting of rock, however, is done from underground.

2.2.1 Consolidation Grouting

Consolidation grouting involves the filling of open joints, separated bedding planes, faulted zones, cavities, and other defects in the rock up to some distance, usually a minimum of one tunnel diameter, beyond the excavation limits. Consolidation grouting strengthens the foundation material and reduces the flow of groundwater into the structure. In the case of high-pressure water shafts and tunnels, consolidation grouting minimizes the flow of water outward through the structure's lining into the surrounding rock after the facility has been put into service. Loss of water to the surrounding rock is objectionable because water is a valuable commodity for human consumption, producing electrical power and irrigation, as well as being used for industrial purposes. Additionally, the water and associated increase in water pressure in the foundation material adjacent to the structure caused by the outward flow may be unacceptable for structural, aesthetic, and environmental reasons.

The defects in the rock surrounding the excavation may be naturally occurring, having existed prior to the excavation, or the defects may have developed as a result of the excavation. It is also possible for existing defects to have been worsened by the excavation process. All excavated surfaces "relax," or move, into the opening after the rock is removed. This movement may cause once tight joints to open and bedding planes to separate. Additionally, the rock beyond the excavation limits is disturbed by forces created by the act of excavating. As a rule, vibrations and expanding gas pressures caused by blasting will disturb the rock surrounding the excavation to a greater degree than does mechanical excavation methods such as using a tunnel-boring machine (TBM) or roadheader excavators.

The requirement to perform consolidation grouting must be included in the contract specifications. The need for consolidation grouting, as determined by the designer, will be based upon geotechnical data that is collected during the geologic site investigation phase of the project and the construction methods used. A predetermined pattern, shown on the contract drawings, is given for hole layout, spacing, and depth. The specifications contain other technical requirements, such as mix design, injection pressures, and refusal criteria.

Generally, unit pricing is used to pay for consolidation grouting. Items such as the number of linear feet of holes drilled, the number of hole hookups, and the quantity of grout mixed and placed by volume, are typical pay items.

2.2.2 Curtain Grouting

Grout curtains are used for underground hydraulic structures that transport and store water, and for underground reservoir structures that store natural gas and petroleum products. Examples of such structures are pressure shafts and tunnels, underground powerhouses, reservoirs, and pumping stations.

Grout curtains for shafts and tunnels are usually installed radially around the entire perimeter of a structure. Grout curtains for underground chambers, such as powerhouses and reservoirs, are installed symmetrically around the perimeter. In shafts and tunnels they reduce the seepage of water downstream of or beyond the curtain. Around powerhouses they limit the inflow of water into the plant. In reservoirs, curtains limit the outflow of liquids and gases being stored. The groundwater around a structure can be naturally occurring or can be at elevated pressures caused by water infiltrating the surrounding rock upstream of the curtain as a result of the underground development, Figs. 2-14 and 2-15 show some typical grout curtain layouts for circular tunnels. In these two figures, primary and secondary holes are shown but tertiary and quartenary holes are left out for drawing clarity.

Fig. 2-16 shows a grout curtain used to reduce the seepage of water into the soil overburden at a tunnel portal. If water were allowed to saturate the soil, the soil could become unstable and lead to slope failure. Fig. 2-17 depicts a grout curtain used to reduce the seepage of water into an underground powerhouse. Water entering the powerhouse could create a safety hazard by causing slippery conditions for workers or posing electrical dangers. It could also cause equipment maintenance problems as well as add costs for additional pumping and continuous dewatering over the life of the powerplant.

2.3 STRUCTURAL GROUTING

Structural grouting is used to improve the ground-structure interface by filling any voids left by the construction process. The filling of voids helps ensure full contact, thus maximizing the transfer of loads between the foundation material and the structure.

2.3.1 Contact Grouting

Contact grouting involves the filling of voids between concrete linings, cast-in-place or precast, and the host geologic material. It also includes the filling of voids behind steel and cast iron liner segments. Additionally, it is used to fill similar voids between steel penstock lining backfill concrete and the host rock. Some excavations in soil and rock require an initial support using precast concrete segments; steel ribs and lagging; or shotcrete followed by a cast-in-place concrete liner. The voids left between these two lining systems may require contact grouting. Contact grouting is sometimes referred to as "backfill" or "backpack" grouting.

GROUTING METHODS

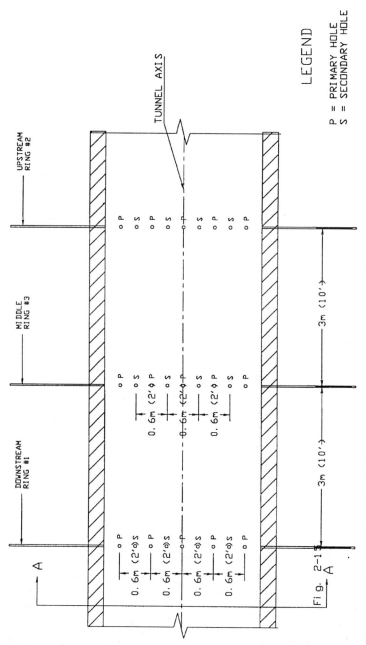

FIG. 2-14. Longitudinal View of Typical Tunnel Grout Curtain Showing Grout Ring and Hole Spacings

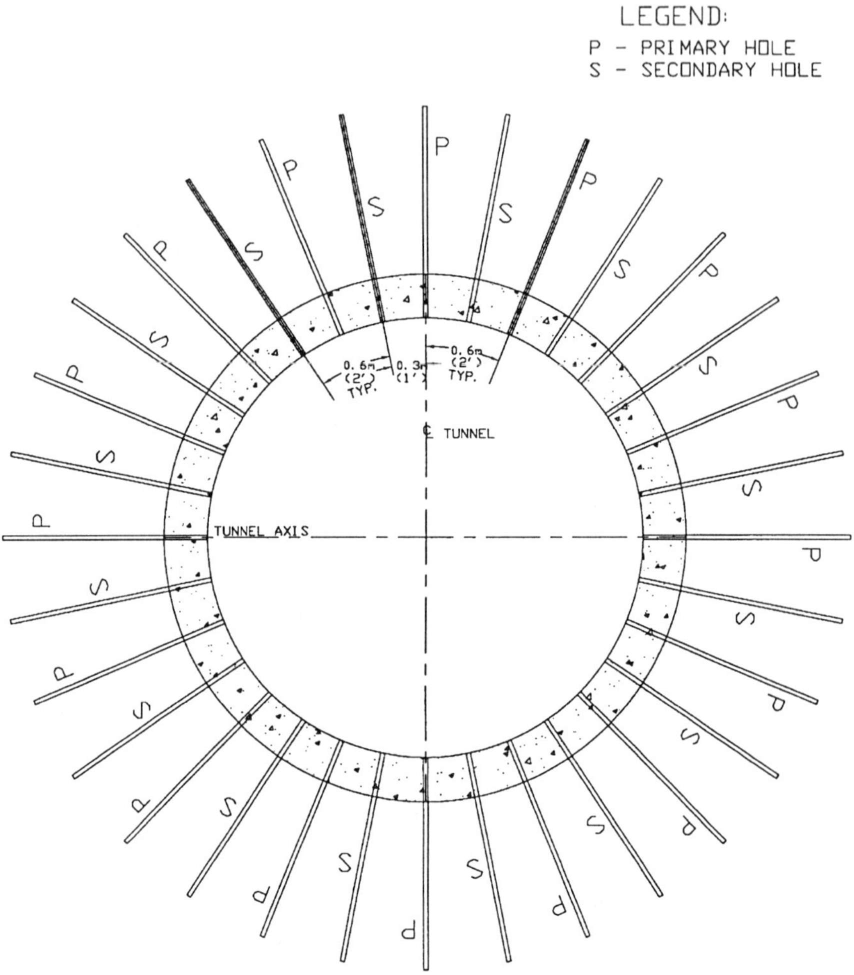

FIG. 2-15. Section A-A from Fig. 2-14; Typical Section of Tunnel Grout Curtain

2.3.1.1 Cast-in-Place Concrete Lining

The occurrence of voids between cast-in-place concrete linings and rock is not uncommon, particularly when concrete is placed overhead, as in tunnel crowns and chamber roofs. Voids can also form in the lower quarter-arcs in circular tunnels containing reinforcing steel and around curved structures such as elbows that form the transition from shafts into tunnels.

Voids in concrete placed overhead usually occur because concrete behaves like a fluid during placement and consolidation by vibration before it

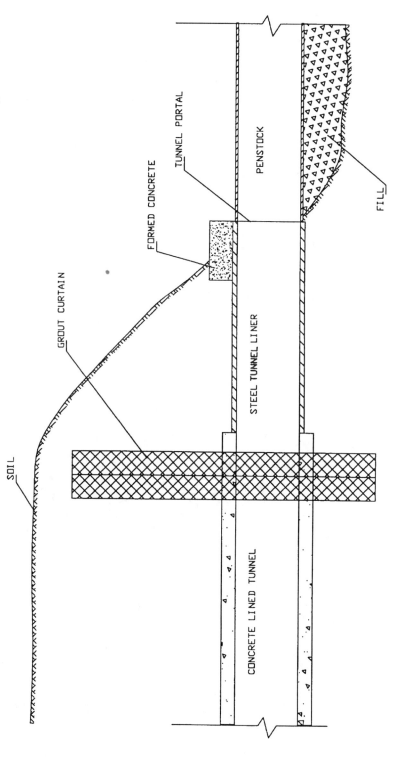

FIG. 2-16. Example of Transition from Concrete-Lined Tunnel to Steel-Lined Tunnel (Penstock) to Aboveground Penstock Showing Location of Grout Curtain

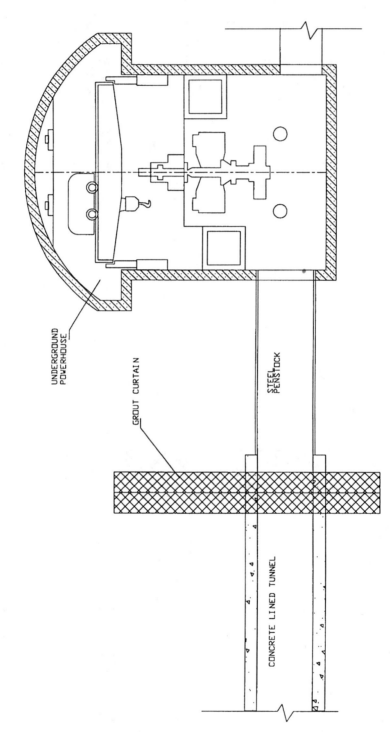

FIG. 2-17. Example of Transition from Concrete-Lined Tunnel to Steel-Lined Penstock Just Upstream of Powerhouse

takes an initial set. In its fluid state, concrete tends to maintain a horizontal surface; therefore, a void will form at the high point of a pour (Fig. 2-18). Voids also develop due to the presence of trapped air, a poor concrete placement procedure, insufficient concrete slump, or unstable concrete (bleed susceptible). Obstructions to concrete flow during placement from items such as rockbolts, mine straps, embedded conduit and piping, or steel sets and lagging, can also cause voids.

The size of a void opening can range from a millimeter (fraction of an inch) to a 0.3 m (1 ft) or more. The areal extent of the void may be localized, covering less than 0.1 m² (1 ft²) or it may extend over tens of square meters (hundreds of square feet).

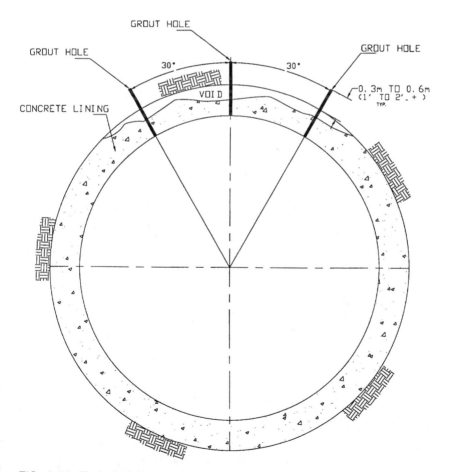

FIG. 2-18. Typical Void and Contact Grout Hole Locations Associated with Cast-in-Place Tunnel Lining

The requirement to perform contact grouting must be included in contract specifications. The need for contact grouting, as determined by the designer, is based on such factors as the operational parameters of the project and the requirements for ground-structure interaction. Contact grouting should be made mandatory by the contract documents when a cast-in-place liner is specified. A predetermined pattern is given for hole layout, spacing, and depth, and is shown on the contract drawings. Two- and three-hole patterns on 3 m (10 ft) centers are common (Figs. 2-18 and 2-19). The other technical requirements, such as mix design, injection pressure, and refusal criteria, are specified. Generally, contact grouting is not paid for separately as a pay item, but is considered incidental to the placement of the cast-in-place and precast liners.

2.3.1.2 Mapping Cracks in Concrete Linings

Most cracks in newly constructed cast-in-place concrete linings are caused by normal concrete shrinkage during the curing process. These cracks create a potential flow path for groundwater to enter the structure. A path is also created by the cracks for high-pressure gases and liquids, stored or transported by the structure, to enter the surrounding foundation material.

Shrinkage cracks are usually relatively tight when good concrete proportioning, placement, and curing practices are followed. If, however, the concrete has a higher-than-necessary water:cement ratio, the concrete has a high cement content, or the concrete is allowed to cure at elevated temperatures, the crack openings may become significant.

Normally, most shrinkage cracks are filled concurrently while performing contact and consolidation grouting operations. For this reason, a simple map or sketch should be made to document the locations and orientations of the cracks before grouting is started. Then, using the maps, a detailed record of water and grout flows emanating from the cracks during grouting should be kept by the grouting inspector. These records need only be qualitative in nature, but should be consistent. Such constancy may prove to be somewhat difficult when grouting is performed on more than one shaft, which makes record keeping by more than one person a necessity. Observations of crack behavior should be recorded in the inspector's field book or directly onto a copy of the crack map. An example of the level of detail required might read, "Radial crack at station 18+03 began making a slight amount of water between 10 o'clock and 2 o'clock, 5 min after contact grouting was started at station 18+17. Only a trace of color at 15 min." Recording this type of information keeps with the philosophy that it is better to have information and not use it than to have no information at all. If not properly documented, this type of information is lost forever with no hope of re-creating it. The crack maps also have a more immediate use during grouting. By carefully monitoring the cracks while grouting is being performed in the immediate

GROUTING METHODS

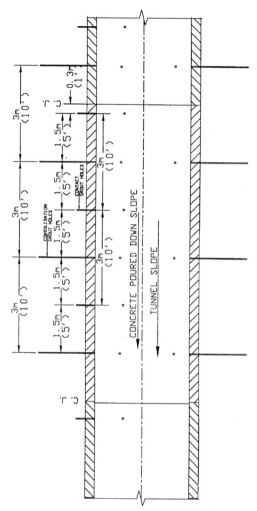

FIG. 2-19. Longitudinal View Illustrating Location of Contact and Consolidation Grout Holes when Tunnel Slope and Cast-in-Place Concrete Tunnel Liner Are Placed in Same Direction

area, say approximately 20 m (60 ft) in any direction, any large grout leaks emanating from the crack can be discovered and plugged. If left unplugged, these leaks can give misleading grout-take data for the contact and consolidation grouting.

Other potential sources of grout leaks, such as construction joints, form-tie holes, and concrete cold joints, should also be mapped and carefully monitored. Holes drilled into the concrete lining to support temporary construction utilities can be particularly troublesome sources of grout leaks.

2.3.1.3 Precast Concrete Lining

The use of precast concrete liners is common in circular tunnels excavated with a TBM in both soft-ground and rock tunnels. They are used as both primary and secondary (final) lining systems. The use of precast concrete as a primary lining is less common in tunnels excavated by the drill-and-blast method, shafts, and underground chambers.

Several precast lining pieces, called "segments," fit together to form a complete circle, referred to as a "ring" (Figs. 2-20 and 2-21). The number of segments required to complete a ring varies, depending on the size of the tunnel and the designer's preference. The segment and ring design giving dimensions, amount of reinforcing steel, and the location of the contact grout

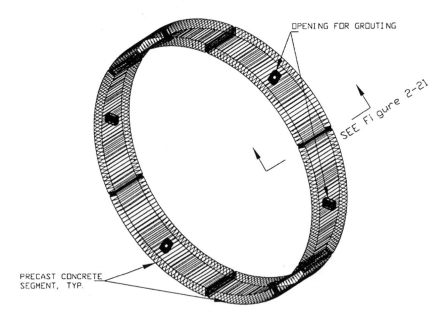

FIG. 2-20. Typical Precast Concrete Tunnel Segment Liner

GROUTING METHODS

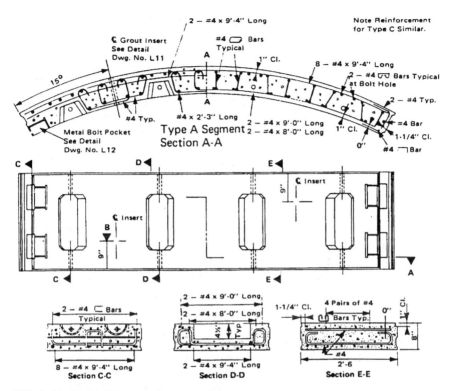

FIG. 2-21. *Typical Precast Concrete Segment*

holes are shown in the contract drawings. The segment design is sometimes made the responsibility of the contractor, but such practice is not widespread.

The contact grout holes, installed in the lining segments by the precast manufacturer during fabrication, often serve a dual function. In addition to acting as grout-injection points, these holes serve as handling holes and lifting points during lining segment erection.

The contact grouting of precast concrete liners serves the same function as the contact grouting of cast-in-place concrete linings, that is, to fill the voids left between the liner and the host geologic material. The genesis of the voids is, however, different. The voids formed in a circular, TBM-excavated rock tunnel using a precast concrete lining tend to be symmetrical and continuous, extending parallel to the tunnel axis. This is because both the TBM tunnel excavation and precast liner rings form almost perfect circles. In rock tunnel applications the precast liner is designed with a slightly smaller outside diameter than the bored tunnel, usually 50–100 mm (2–4 in.), to allow for erection. The locations and extent of the voids are thus more predictable for precast liners than for cast-in-place liners.

The size, location, and areal extent of the voids depend on such factors as

- The difference between the excavated diameter of the TBM tunnel and the outside diameter of an erected precast concrete ring
- The fabrication and erection tolerances for the precast segments
- The degree of horizontal and vertical curvature of the tunnel
- The amount of deviation in line and grade of the excavated tunnel
- The amount of overbreak or rock fallout beyond the excavation limits that results from the excavation crossing geologic discontinuities in a hard-rock application

Since these voids are expected, and their size and locations are predictable with some degree of certainty (except for the ones caused by geologic conditions), the segments for rock tunnels are fabricated with the contact grout holes in predetermined locations. The grout holes are installed symmetrically around the ring, with one or two holes in each segment.

The void-filling procedure differs somewhat between cast-in-place and precast linings. Typically, shortly after a precast ring erecting is completed, the space between the rock and liner is filled with 10 mm (3/8 in.) pea gravel in the hard-rock application. The pea gravel is blown in pneumatically through the grout holes, often using shotcrete placement equipment. The pea gravel helps to keep the ring stable by preventing movement due to construction loading. By using pea gravel, the major portion of the void is filled with gravel, rather than more expensive contact grout. Contact grouting is then used to fill the voids between the individual gravel particles. The contact grouting can be performed a considerable distance behind the excavation and lining erection operations. It is sometimes performed only after excavation and lining operations have been totally completed. The timing of the grout application depends on the size and length of the tunnel and the contractor's scheduling of the work sequence. Cellular concrete can also be used to fill these voids. A more detailed discussion of cellular concrete is given in Section 2.3.4.

In soft-ground tunnel applications, when a precast lining system is used as primary ground support, contact grouting behind the liner is performed as soon as possible after the segments are erected. This helps lessen any inward ground movement caused by the excavation process. If this inward ground movement is not controlled, it can translate upward to cause surface settlement. In urban and industrialized areas, even minimal surface settlement can cause damage to buried utilities, roadways, rail lines, and building foundations.

During excavation, a void is left between the skin of the soft-ground tunnel shield main body and the surrounding ground. The void is caused by intentional overcut or overexcavation by the machine, pitch and yaw of the machine during steering, and the rear seal assembly of the machine. The overcut is created by the machine having a slightly larger diameter at the leading edge of the shield than at its main body. This overcut reduces skin friction between the main body and the surrounding ground. This helps reduce the shoving forces necessary to advance the machine, thus reducing the loads applied to the segments. This also helps make steering of the machine easier. The overcutting usually is between 15 and 20 mm (5/8 and 1 in.) in radius. There is also another large void formed at the rear of the machine tail skin; it is created by the pitch and yaw of the machine during steering to adjust the line and grade. An additional void is created at the extreme rear of the tail shield by the rear seal assembly.

The total void between the segments and the soil ranges from 50 to 150 mm (2 to 6 in.). The size of the void depends on such factors as the overcut of the machine, the machine tail-skin thickness, and the rear seal assembly dimensions.

In an effort to fill these voids, contact grout is injected between the precast segments and the surrounding soil. The grout can be placed through grout holes cast into the segments or through the tail shield of earth pressure balance (EPB) machines or slurry shield soft-ground tunneling machines. When using these types of tunneling machines, the grout is injected behind the liner while the liner is still in the tail shield.

The tail skin of the machine is equipped with two or three sets of steel-brush or rubber seals. The steel-brush-type seals are more common. The chambers between the seals are filled with biodegradable grease to help maximize the seal effectiveness. The chambers are continuously supplied with grease via grease lines installed in the tunnel shield. These types of sealing systems allow for pressure injection of grout behind the last segment ring erected while preventing the grout from travelling forward and entering the main body of the shield.

When geologic conditions allow the use of an open-faced shield tunneling machine, precast segments are often used as the primary ground support. Precast segments are sometimes used both for the primary and final lining systems. When only one liner is used, this is referred to as a "one pass lining" system. When precast segments are used with an open-faced shield tunneling machine, the segments are assembled within the tail shield of the machine. As they leave the tail shield, the segment ring is expanded outward to make contact with the surrounding soil. The ring is expanded using jacks or a wedge-shaped segment closure piece. In this application contact grouting behind the precast lining is often still necessary. However, the grout takes behind such liner systems are generally low because the liner is theoretically in full contact with the surrounding ground. The contact grouting is usually concentrated in the crown of the tunnel. Grouting is performed to compensate for any ground loss above the tunnel crown that may have occurred before or during liner erection. It can be argued that in this application, the grouting is actually performed more as a form of compaction grouting rather than contact grouting.

Steel sets and wood lagging are also used as primary lining systems with open-faced shield tunneling machines. In these cases, precast segments are often used as the secondary or final lining systems. The annulus between the segments and the steel ribs and lagging are backfilled or contact-grouted using holes cast into the segments. The annulus is often first filled with pea gravel as an economic measure, with subsequent contact grouting used to fill the voids in the gravel. Cellular concrete is also used as a backfill material in these applications.

2.3.1.4 Steel and Cast Iron Segment Linings

Steel and cast iron segment lining systems are used in both soft-ground and rock shaft and tunnel applications. They can be used as either primary or secondary (final) lining systems. They sometimes serve dual functions.

Several steel or cast iron pieces, called segments, fit together to form a complete circle referred to as a "ring." In addition to circular shafts and tunnels, segments are also used to line elliptical shafts and horseshoe-shaped tunnels, as well as other various shapes of underground structures. The number of segments required to complete a ring varies, depending on the size and shape of the structure and designer preference. Each segment piece is fitted with one or two contact grout holes that are built into the segments during fabrication. These holes are generally fitted with threaded pipe unions shop-welded or cast onto the inside face of the segment skin. They can also be drilled and tapped. The threaded holes are usually between 150 and 200 mm (1.5 and 2 in.) in diameter. Holes should be kept sealed with pipe plugs until the pipe nipple is installed just prior to grouting. Individual segment pieces within a ring are bolted to each other. Segments are also bolted from ring to ring using bolt holes drilled or punched in all four of the flanges of each individual segment piece. Steel segments can be manufactured as pressed steel called "liner plates" or fabricated steel made of rolled steel plates with welded stiffeners. Pressed steel liner plates are used most often as primary ground support for soft-ground applications.

Tunnels often require both a primary and secondary liner. In some cases only one liner serves both functions. This is referred to as a "one-pass" lining system. Liner plates are sometimes used in a one-pass system as the final lining for utility tunnels, access portals, and underground passageways. When liner plates are used as the primary liner in a two-liner system, the second or final liner is usually cast-in-place, precast concrete, or steel pipe. This is referred to as a "two-pass" lining system.

GROUTING METHODS

Soft-ground tunnels driven with shields always require some form of primary liner system for ground support and to react the shoving force of the shield jacking system. This primary support can be supplied by precast concrete segments, steel liner plates, fabricated steel segments, cast iron segments, or steel ribs and lagging.

When the loads imposed on a lining plate system are high, steel sets are used to help support and reinforce the liner plates. These loads can be imposed by the surrounding soil and transportation systems and other structures located both above or below the ground surface in the area of the underground excavation. Fabricated steel segment liners, made from rolled steel plate with welded stiffeners, are used when loads which exceed the capacity of a liner plate system are expected. They are also used when the steel segment liner will serve as the final lining. In this case the steel is usually shop-coated with a coal-tar–based material or painted. A fabricated steel segment liner may also serve as both the primary and final lining in a one-pass system.

Cast iron segment lining systems are commonly used in soft-ground applications when the liner will serve as both the primary and final liner. Since cast iron is resistant to corrosion, cast iron lining systems can be left in a bare metal or uncoated state.

The flanged surfaces of both cast iron and fabricated steel liner segments are machined to help provide a better, tighter fit between segments. Additionally, the segment flanges can be grooved on all four sides to allow the installation of gasket material. A groove-and-gasket joint arrangement is also used in liner plates. The use of gaskets in joints helps to increase the water tightness of the liner system. The gasket also provides a joint seal during contact grouting and they are further used to seal out gases, such as hydrogen sulfide and methane, a common practice in transit tunnels. The size and shape of the groove and the type of the gasket material depends on the quantity, pressure, and chemical composition of the groundwater and gases.

The maximum grouting pressure applied to the segment depends on the design of the segment, the ground conditions, groundwater pressure, and the proximity of nearby structures. Generally, grouting pressures range from 0.3 to 2 bars (5 to 30 psi). The Commercial Pantex Sika Incorporated recommends a range of 0.3 to 0.7 bars (5 to 10 psi) for contact grouting behind their liner plates.

The maximum allowable grouting pressure to be used for grouting behind a liner system must be given in the contract documents. The designer should always specify the highest pressure that can be safely applied based on a structural analysis of the liner system. The contractor should always apply the maximum pressure allowed during grouting. This helps ensure the complete filling of all voids with grout.

2.3.1.5 Steel Penstock Lining

Underground structures which serve as conduits to transport water at elevated pressures are usually steel lined for part or all of their length. Fig. 2-22 shows a 4.3 m (14 ft) diameter steel penstock for a high-head hydroelectric development project. Pressure shafts and tunnels associated with hydroelectric developments and water conveyance systems are good examples.

Pressure shafts and tunnels are classified as low-, medium-, or high-pressure based on the internal operating pressures expressed in meters or feet of head. As a general guideline, low head is less than 30 m (100 ft), medium head is 30–150 m (100–500 ft), and high head is greater than 150 m (500 ft). These pressure classifications vary over a wide range from one agency to another and one geographical region to another.

The steel-lining wall thickness and the importance of the interactions at the interfaces among the foundation material, backfill concrete, and the steel lining, all increase with increasing operating pressure. Therefore, the importance of the quality of the contact grouting program also increases with increasing operating pressures.

The excavation limits of shafts and tunnels in which steel linings are to be installed are generally specified to be 0.6–1.2 m (2–4 ft) larger than the outside diameter of the lining. This overexcavation allows for the installation

FIG. 2-22. Steel Penstock of 4.3 m (14 ft) Diameter for High-Head Hydroelectric Development

and alignment of the liner. The overexcavated space between the liner and the rock or primary liner must later be filled with plain or reinforced concrete. In this application, steel lining also serves as a stay-in-place concrete form for the concrete, thus performing the same function as the removable formwork used for cast-in-place concrete linings. The resulting concrete/rock or primary liner interface is subjected to the same void causing mechanism that occurs when using cast-in-place concrete tunnel linings, as discussed earlier, in the section on cast-in-place concrete linings. The same contact grout placement procedures are also used to fill the voids in both cases.

One noticeable feature unique to the contact grouting of steel linings is the use of preinstalled internally threaded grout connections that are shop welded to the liner during fabrication (Fig. 2-23). The inside diameter of the threaded grout connection must be 25–38 mm (1–1.5 in.) larger in diameter than the grout hole that will be drilled through the backfill concrete into the rock. This oversizing of the grout connection is done to prevent damage to the internal threads during the drilling operation.

2.3.1.6 Tunnel and Station Linings Using Waterproofing Systems

A recent trend, particularly in transportation projects in the United States, is to incorporate a membrane waterproofing system into the tunnel and underground station structural lining design. A popular waterproofing method used in soft-ground and rock applications is the PVC membrane lining system.

Membrane liners are used in rock excavations performed by the drill-and-blast, TBM, roadheader, and other mechanical excavation methods. After excavation and rock bolting, the rock surface is shotcreted. The shotcrete seals the rock to prevent moisture loss and rock deterioration. It is also used to help create a smooth surface to receive the membrane liner. The use of membrane liners is also common in soft-ground excavations performed by the New Austrian Tunneling Method (NATM).

When a membrane waterproof lining system is used in conjunction with a cast-in-place final liner, contact grouting between the waterproof system and the cast-in-place concrete lining is usually required. In this application the installation of contact grout holes to the cast-in-place concrete lining requires special attention to protect the integrity of the waterproofing system. To be effective the membrane waterproofing system must be completely watertight. To ensure watertightness all of the seams and joints of the PVC membrane system are thermal welded and pressure tested before the final concrete liner is placed against it.

The drilling of contact grout holes through the concrete liner is not possible because of the potential for damaging the waterproofing. Therefore, grout pipes are installed onto the formwork and cast into the concrete during

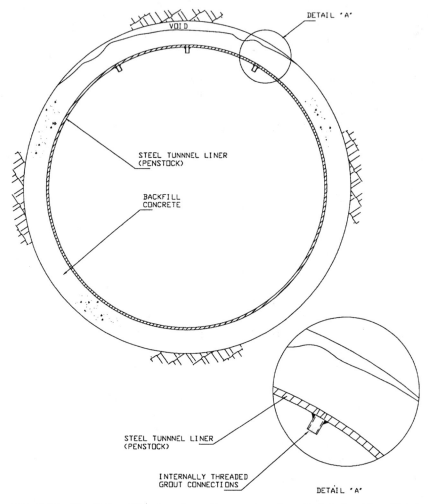

FIG. 2-23. *Steel Liner Showing Grout Connections Installed during Liner Fabrication*

placement. The installation and securing of the grout pipes against the membrane liner is made difficult because no mechanical fastening of the pipes to the liner is possible due to the likelihood of damage to the waterproofing.

A grout-pipe installation practice is to tape the end of a grout to the membrane using standard duct tape. This creates a seal against concrete entering and plugging a grout pipe. During grouting the grout pressure breaks the seal of the tape and allows contact grouting to proceed. Another method of creating a seal is to install a slightly smaller diameter pipe inside

GROUTING METHODS

the grout pipe. The smaller pipe is placed in hard contact with the membrane lining and is mechanically secured in place with the larger pipe. The smaller pipe creates the seal, thereby preventing concrete from entering the larger pipe. The smaller pipe is removed after the concrete has set, thus providing an open pathway for future contact grouting.

In both these methods there is potential for the grout pipes to become dislodged and shift location, thus breaking the seal during concrete placement. Once dislodged, the pipes often become plugged with concrete, making them unusable for grouting. Since drilling a pipe out in an attempt to clean it is not possible because of the potential damage to the membrane liner, the contact grout hole is lost.

In an effort to better secure the grout pipe against movement during concrete placement, the author has devised a method for anchoring grout pipes in place at the end of the pipe, against the membrane. Fig. 2-24 shows a typical membrane waterproofing/concrete-lining system arrangement. The proposed grout pipe anchoring system consists of a watertight PVC insert installed through the membrane liner into the foundation material. While the watertightness of the liner is destroyed by drilling through the liner, it is restored by thermal welding the circular collar of the insert to the membrane liner. Figs. 2-25 and 2-26 show a photograph and sketch, respectively, of the proposed grout pipe anchoring insert.

The 50 mm (2 in.) inside diameter PVC sleeve of the insert extends approximately 400 mm (8 in.) into the foundation material beyond the

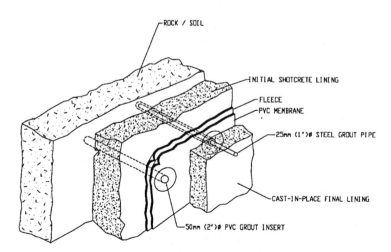

FIG. 2-24. Typical Membrane Waterproofing/Concrete-Lining System Arrangement Showing Grout Pipe Inserts (Drawn by Dr. G. Sauers Corp. 1995)

FIG. 2-25. Prototypical Grout Pipe Insert

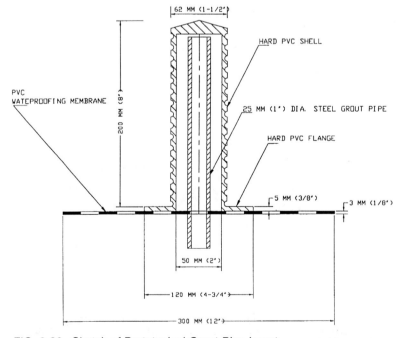

FIG. 2-26. Sketch of Prototypical Grout Pipe Insert

excavation limits. During installation, the grout pipe is installed approximately 150–175 mm (6–7 in.) into the sleeve. This isolates the exit end of the grout pipe well above the concrete during placement. When grouting, the grout exits the pipe, flows down the annulus between the pipe and the sleeve, and fills any voids between the membrane and the concrete liner.

2.3.2 Embedment Grouting

Embedment grouting entails the grouting of voids between items embedded in the concrete and the concrete surrounding the item. Some examples of common embedded items are penstocks, large-diameter pipe bends and elbows, electrical conduit, small utility piping, structural shear lugs, stiffener plates, and equipment bearing pads. In the case of penstocks, embedment grouting is used to fill voids between the penstock and the backfill concrete. Any voids between the backfill concrete and the foundation material are filled during contact grouting.

The voids requiring embedment grouting may be the result of concrete shrinkage, air being trapped during concrete placement, or poor concrete placement methods. The size and shape of the embedment can also restrict the flow of concrete during placement, and the physical shape or geometry of the concrete pour is also a major factor in the formation of a void. This is particularly true in steel-lined shafts and circular tunnels, at turns and bends in the shaft and tunnel alignment, diameter changes, intersections, and so on.

The need to perform embedment grouting must be included in project specifications. The maximum areal extent or the maximum void surface area which will be allowed before embedment grouting is required must be specified. The need for embedment grouting, as determined by the designer, is based on such factors as the operational parameters of the project, the structural requirements of the embedment, and the recommendations of the embedment manufacturer. Embedment grouting should not commence any earlier than 28 days after the placement of the concrete so as to allow for maximum concrete shrinkage prior to the grouting application. Embedment grouting should also be the last grouting operation performed, following, in order, contact, consolidation, and curtain grouting. These methods fill many of the voids around an embedment, thus reducing the amount of embedment grout needed. The void around an embedment may range in thickness from less than 2 mm (1/16 in.) to tens of millimeters. The areal extent may be small, a matter of square millimeters, or it may extend over several square meters.

Portland-cement–based grouts are normally specified for embedment grouting. Ultrafine cements are sometimes specified where the thickness of the void is expected to be very small, on the order of 2 mm (1/16 in.) or less. Ultrafine cement is more commonly required when voids between backfill concrete and steel tunnel linings need to be filled, as in penstocks used for pressure tunnels. Embedment grouting becomes necessary when the interaction between the concrete, embedment, and rock is a critical structural requirement.

A procedure that locates and determines the extent of voids around embedments is sounding the embedment with a hammer or steel rod. In the case of a steel tunnel liner or penstock, sounding is used to locate the perimeter of the void. The sound of steel striking steel will vary from a sharp ring, in the case of no void, to a "drummy" or hollow sound, when there is a void behind the steel. A method for outlining the extent of the void is to spray paint a dot at the location of the interface between the void and solid material. The dots are then connected, thereby providing the outline of the void (Fig. 2-27).

A minimum of two grout holes are required for each void: one to inject the grout and one to allow air to vent (Figs. 2-28 and 2-29). Several sets of injection and vent holes may be required to grout a single void depending on the void size and shape. The injection holes must always be at the lowest point of the void, whereas the vent holes must be at the highest point (Fig. 2-29).

A 10–13 mm (3/8 to 1/2 in.) diameter grout and vent hole is adequate. The hole need only be drilled through the embedment itself using a high-quality steel drill bit. A masonry drill bit is then used to extend the hole approximately 13 mm (2 in.) into the backfill concrete. This is done to allow room for a pipe-tapping tool to be used to cut pipe threads into the embedment. The threads are needed to accommodate the pipe nipple used for grouting.

As with other types of grouting, the general locations where embedment grouting may be necessary must be called out in contract drawings and specifications. The number of holes and quantities for grout injected, however, may vary widely. For this reason embedment grouting is normally paid for on a unit price basis. Items such as the number of holes drilled, the number of hole hookups, and the quantity of grout mixed and placed by volume are typical pay items.

2.3.3 Prestressing

Prestressed grouting is used with cast-in-place concrete shaft and tunnel lining when they are subject to elevated internal operating water pressure. Prestressed grouting of cast-in-place concrete lining can be an economically superior alternative to using steel liners (Gonano and Sharp 1984). This method is not, however, widely used at present and is performed by a limited number of engineering and specialty contracting firms. For these reasons only an overview of the method is presented here.

The use of such lining provides the most hydraulically efficient means to transport water. The lining also minimizes or eliminates water loss from the conduit into the surrounding rock. Water loss is objectionable because of the

GROUTING METHODS

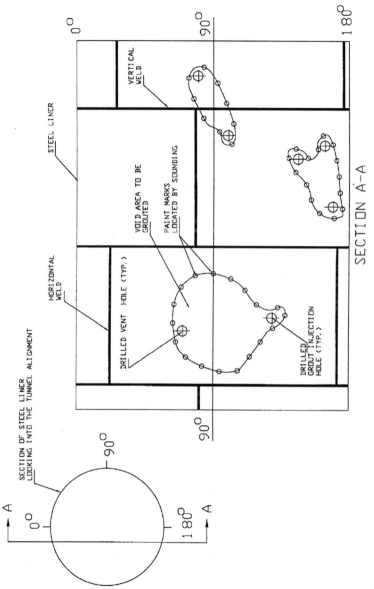

FIG. 2-27. *Longitudinal View of Steel Tunnel Liner Showing Void Areas Located and Marked with Paint*

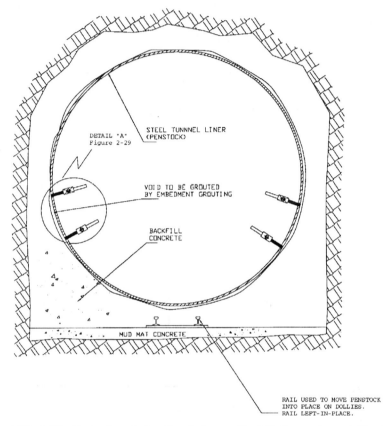

FIG. 2-28. Embedment Grouting between Backfill Concrete and Steel Tunnel Liner (Penstock)

economic waste of the water resource itself. Also, the exfiltrating seepage forces may increase groundwater pressure in the rock surrounding the structure to an unacceptable level.

Conventional cast-in-place concrete shaft and tunnel linings, that is, linings without prestressing, continue to be used extensively in hydroelectric and pumped storage developments. The current trend, however, is to design these facilities with higher internal water pressure coupled with development schemes that are located in more marginal geological conditions than previously encountered. This necessitates that the lining contribute to the structural stability of the conduit. To accomplish this, steel linings are commonly used, as discussed in the subsection on steel penstock lining. Steel linings are, however, more costly than cast-in-place concrete lining. Prestressing thus may offer an economical alternative to steel lining.

GROUTING METHODS

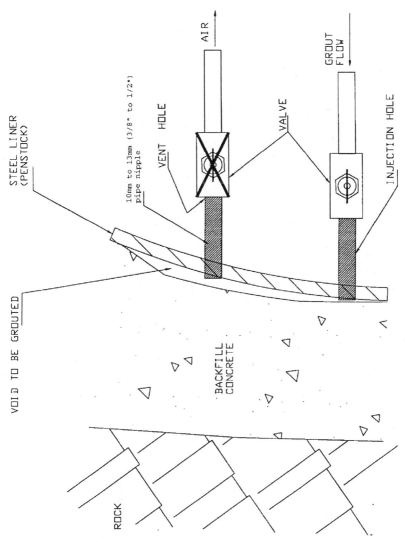

FIG. 2-29. Detail of Grouting and Venting Holes Used for Embedment Grouting of Steel Tunnel Liner

The concept of prestressing involves providing equally distributed external forces from outside the lining that oppose the forces of the water that act from within during operation (Fig. 2-30). The external forces acting inward produced by the prestress will prevent the excessive outward deflection of the lining, thus precluding the development of tension cracks through which water can escape.

Like contact grouting, prestress grouting is applied to the space between the lining and the rock. Unlike contact grouting, however, prestress grouting is applied symmetrically around the shaft or tunnel circumference. It is injected into an intentionally created space under high pressures that are on the order of 5–7 Mpa (750–1,000 psi). The intentionally created space is formed by installing a bond breaker to the rock surface prior to concrete placement. Because of the high pressure, normal grouting packers and pipe nipples are not used. Longitudinal or radial grout pipes are used instead (Fig. 2-31).

Prestress grouting will deflect concrete lining inward. This movement, called convergence, is measured, with tape extensometers in conjunction with fixed convergency measuring points, during and after grouting application to ensure that design assumptions are met. Other geotechnical instru-

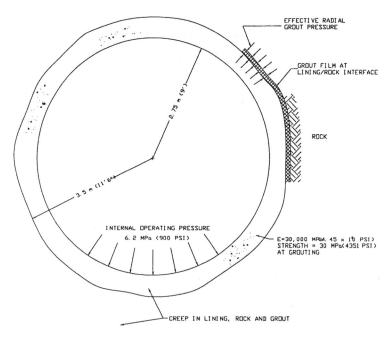

FIG. 2-30. Tunnel Design Model for Calculation of Prestressing Requirements (Gonano and Sharp 1984)

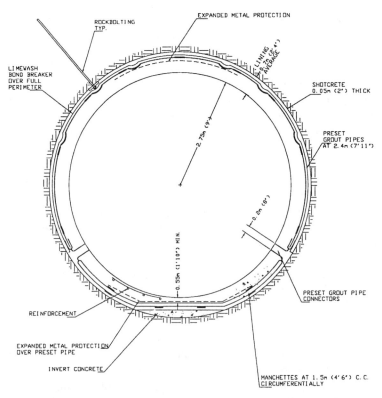

FIG. 2-31. *Cross-Sectional View of Grouting Fixtures for Prestressing Grouting (Gonano and Sharp 1984)*

mentation, such as load cells, borehole extensometers, and inclinometers, may be used. The measurements are compared to acceptance criteria that are theoretical deflections developed by calculation.

2.3.4 Cellular Concrete

Although the term "cellular concrete" might imply a conventional aggregate and cement mixture, it actually applies to a cementitious grout and foaming agent mixture, which may or may not contain fine aggregate. This grout and foaming agent mixture produces a low-density, nonshrink, economical material used for bulk backfilling. The term cellular concrete is used in the mining and underground construction industries to denote a material used for backfilling large voids created by mining operations and the filling of the annulus between the primary and the final tunnel linings. It is also used in foundation and soil modification applications.

The American Concrete Institute (ACI) defines cellular concrete as, "a lightweight product consisting of portland cement, cement silica, cement-pozzolan, lime-pozzolan, or lime-silica pastes, or pastes containing blends of these ingredients and having a homogeneous void or cell structure, attained with gas-forming chemicals or foaming agents (for cellular concrete containing binder ingredients other than or in addition to portland cement, autoclave curing is usually employed)" (ACI 1967). In cellular concrete the density control is achieved by substituting microscopic air cells for all or part of the fine aggregate.

ACI has also published a "Guide for Cast-in-Place Low-Density Concrete" (*ACI 523.1R-86*), which provides information on materials, properties, design, and the proper handling of cast-in-place concretes having oven-dry unit weights of 88 kg/m^3 (5.5 lb/cu ft) or less. ASTM has written several standard specifications that apply specifically to cellular concrete and foaming agents including these:

- *ASTM C869-80.* "Foaming Agents Used in Making Performed Foam for Cellular Concrete"
- *ASTM C796-80.* "Standard Method of Testing Foaming Agents for Use in Producing Cellular Concrete Using Performed Foam"

The low-density properties of cellular concrete are achieved by using foaming agents which can produce a backfill material with an air content of up to 80%. Developed in the early 1980s, cellular concrete was first called foamed cement. It was originally used in Europe in the long-wall coal mining industry to help solve roof control and caving problems. The cellular concrete helped replace the need for timber cribbing, which is labor intensive and dangerous to install, and furthermore can pose a fire hazard. Fosroc Inc., which helped develop the technique in Europe, expanded their market to the United States. Their U.S. subsidiary, Celtite Inc., produced TEKFOAM II, a more durable and higher strength foaming agent. Through the joint cooperation with the Mine Safety and Health Administration and the United States Bureau of Mines, Celtite began a testing program in 1990. The program was designed to demonstrate the suitability of a foamed cement for use as permanent ventilation seals in coal mines in the United States. Since then there have been hundreds of these seals placed in U.S. coal mines using Celtite's TEKSEAL II foamed cement.

The Mearl Corporation has also developed cellular concrete products. Mearl GEOCELL, widely used in both the mining and tunneling industries, is a lightweight cellular concrete which is batched on or off the jobsite. This is done by mixing a portland cement slurry, with or without fine aggregate, and Mearl GEOFOAM Liquid Concentrate preformed foam. By using the proper mix design and controlling the mixing process, a cellular concrete can be produced to meet a range of specific application requirements. A typical range of densities for Mearl cellular concrete is from 318 to 1,920 kg/m^3 (20 to 120 lb/ft^3) with compressive strengths ranging from 0.17 to 17.5 Mpa (25 to 2,500 psi). Mearl GEOCELL is used for the following mining industry applications:

- Subsidence abatement
- Fireproofing
- Mine fire abatement
- Ventilation seals
- Arch backfill
- Roof cavities or void filling

Mearl GEOCELL is also used in the underground and civil construction industries as a lightweight backfill material for tunnel linings and for solving shallow surface soil stabilization problems. Mearl GEOCELL was notably used as a controlled low-strength material to replace conventional fill at the Boston Commonwealth Pier. Approximately 60,000 m^3 (78,000 cu yd) of Mearl GEOCELL, containing 80% air by volume, was used to stabilize the pier fill, and to relieve the load on the compressible subsoil and walls by replacement of the existing fill (Mearl Corp. 1992a).

Mearl GEOCELL product was also used on the North Outfall Replacement Sewer (NORS) Project in Los Angeles as a tunnel lining backfill. GEOCELL was used to fill the 250 mm (10 in.) annulus between the primary precast segmental liner and the final lining consisting of prestressed concrete cylinder pipe (PCCP). A record 64,200 m^3 (84,000 yd^3) of cellular concrete was mixed and placed at rates up to 80 m^3/h (105 yd^3/h). It was pumped to distances of up to 3,048 m (10,000 ft). The cellular concrete grout slurry was designed to develop a minimum compressive strength of approximately 900 kPa (130 psi). The grout slurry was, for the most part, batched off site and delivered to the shaft site by ready-mix concrete trucks. A custom-designed high-output mixing and pumping plant was supplied by the Mearl Corporation and Pacific International Grout Company. The mixer and pumping plant, located at the bottom of the shaft, was used to remix and agitate the slurry grout. The mixture was then pumped through a 100 mm (4 in.) diameter slick line using a progressing helical cavity pump. The specialized foam injection/mixing plant was located within the PCCP near the point of injection. After the Mearl GEOFOAM Liquid Concentrate had been added to the grout mix, the Mearl GEOCELL was pumped into the space between the primary precast concrete segmental liner and the PCCP. The delivery line was connected to grout nipples installed into threaded holes cast into the PCCP at the spring line. Upon completion of the grouting, the grout holes and PCCP joints were patched using a dry-patch sand-cement mixture (Mearl Corp. 1992b).

2.4 SHAFT GROUTING

Like tunnel and chamber grouting, shaft grouting is performed to reduce seepage and to modify or strengthen in situ foundation material prior to excavation. However, reducing seepage or controlling the inflow of groundwater is the most common reason for performing shaft grouting. Groundwater inflows can be particularly troublesome in shafts because, as the shaft excavation is advanced, the inflows from individual water-bearing geologic formations intersected by the shaft excavation, are cumulative at the shaft bottom. During shaft excavations using the drill-and-blast method, the shaft bottom is the exclusive work area for the drilling, explosive loading, blast wiring, and mucking operations. To make these operations cost effective and safe, the water must be pumped from the work area to the surface. As the volume of water and shaft depth increase, pumping costs can become prohibitive. There is also a point when the volume of water becomes so great, or the shafts so deep, as to require pumping equipment larger than practical to allow for economic shaft excavation. Inflows of groundwater can also make shaft linings difficult and costly to install. These water inflows, if not properly controlled, can cause a negative impact on the quality of shaft linings. For example, inflow can wash out the cement paste from cast-in-place concrete linings, leaving only uncemented pockets of aggregate. Also, if grouting is not performed to reduce inflows during the construction stage of the project, the inflows may continue to be a costly dewatering and maintenance problem during the life of the facility. It is for these reasons that shaft grouting is performed prior to starting or during the shaft excavation to eliminate or greatly reduce the amount of water entering the shaft.

Shaft grouting can be performed from the surface or shaft collar prior to beginning excavation; this is called pregrouting. Shaft grouting can also be performed from within the shaft at various levels, called grout covers, during excavation. Grouting from the surface has the advantage of allowing grouting to be performed before shaft excavation begins. Therefore, by using good preplanning, shaft grouting from the surface can often be accomplished during the early stages of the project. This pregrouting helps speed up shaft excavation and lining activities, which are usually on the critical path of the project. Shaft excavation and lining installation are also some of the most expensive components of an underground development (MacGillivray 1979; Nel 1981). Therefore, any steps to minimize the duration of their construction will reduce the overall project cost. The decision to grout from the surface versus from within the shaft depends on the overall depth of the shaft as well as the depth, orientation, and number of water-bearing geologic formations or zones which require grouting. A combination of pregrouting from the surface and grouting from within the shaft may be used, depending on site-specific requirements. Both pregrouting and grouting from within the shaft are performed using either the stage-down or stage-up grouting methods. The stage-down method is used more widely. In a typical stage-down method the grout hole is advanced in 3–6 m (10–20 ft) stages, a packer is installed and the stage is grouted. After time to allow the grout to set, the hole is advanced another 3–6 m (10–20 ft) and grouted. This procedure is followed until the full depth of the grout hole is reached. The procedure should be modified as needed to better suit the specific geologic conditions of the shaft. If, for example, one or several isolated water-bearing formations require grouting, the drill hole should be advanced directly to these zones and grouted using a single or double packer arrangement. The methods, procedures, and equipment used for shaft grouting as they pertain to proportioning, mixing, and injection are the same as those described in the sections grouting in soil and grouting in rock.

2.4.1 Pregrouting of Shaft from Ground Surface

The first step in pregrouting from the surface is to construct a shaft collar. The shaft collar should be constructed of reinforced concrete approximately 0.4–0.6 m (1.5–2 ft) thick. The shaft collar serves several functions during the shaft grouting, excavation, and lining operations. First, it provides a level and structurally sound working surface allowing heavy equipment to be set up around the top of the shaft. The collar also acts as a dam, preventing runoff from rain and melting snow from entering the shaft. To be effective as a dam, the shaft collar should be constructed at a slightly higher elevation than the surrounding ground. During pregrouting the shaft collar provides a good working platform for the drilling, mixing, and grout-placement operations. Figs. 2-32 and 2-33 show a typical shaft collar arrangement. Shaft collars are used for shafts which are started or collared in both soil and rock. To aid in construction, a shaft collar should be installed for all shaft construction, whether or not pregrouting is used. When a shaft is collared in rock the first 3–6 m (10–20 ft) of rock directly beneath the collar are consolidation-grouted using methods described in the section on consolidation grouting. The consolidation grouting fills open joints and fractures, which are common in rock formations near the ground surface. This helps prevent the loss of grout up through these openings during the production grouting of the lower stages. It also aids in stopping surface water from entering the shaft through open joints. If a shaft is collared in soil and transitions into rock, or when the shaft is excavated in soil to its full depth, grouting methods described in the section on grouting in soil are used.

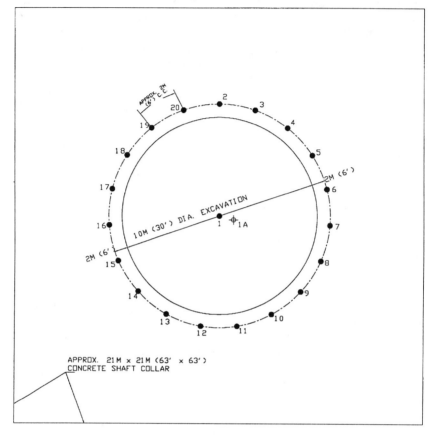

HOLES DRILLED AND GROUTED IN THE FOLLOWING SEQUENCE:
1
2, 4, 6, 8, 10, 12, 14, 16, 18, 20
3, 5, 7, 9, 11, 13, 15, 17, 19

FIG. 2-32. *Plan View of Typical Pregrouting Hole Layout of 10 m (30 ft) Diameter Shaft*

Fig. 2-32 shows a typical pattern for pregrouting hole layout and the grouting sequence, and Fig. 2-33 shows a section view. The actual number of holes and their spacing depends on the diameter of the shaft and the permeability and structural characteristics of the formation(s). The number 1 hole in Fig. 2-32 should be drilled during the site geotechnical investigation phase of the project. Data obtained during the investigation is used to design the shaft grouting program. The number 1 hole is grouted first during

GROUTING METHODS

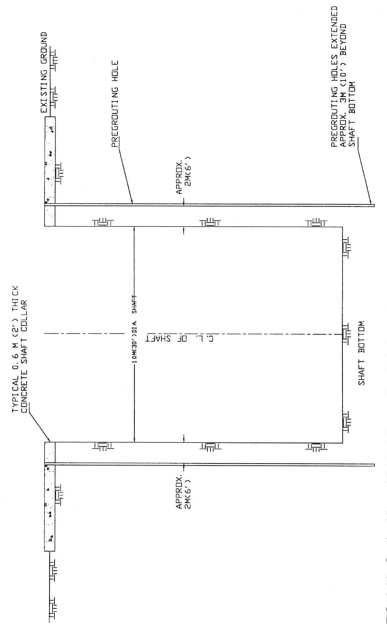

FIG. 2-33. Section View of Typical Pregrouting Hole Layout

pregrouting. The hole numbered 1A in Fig. 2-32 is optional. It is drilled only after the last pregrouting hole has been grouted. It is used to conduct permeability testing, the results of which are used to evaluate the effectiveness of the pregrouting program. If the test results are not within acceptable limits it will be necessary to add additional pregrouting holes. The additional holes should be split-spaced or placed between the holes of the first pregrout ring. This process is continued until the desired reduction in permeability is achieved.

2.4.2 In-Shaft Grouting

When shaft depths increase beyond several hundred meters or when water-bearing geologic formations, that require grouting, are known to exist at isolated locations relatively deep in the shaft, it may become cost effective to perform in-shaft grouting. When in-shaft grouting methods are used the shaft excavation must be temporarily stopped while the drilling and grouting operations are performed. The grout hole depths can range from tens of meters to several hundred meters. The hole layout and grouting sequence is basically the same as those used for pregrouting as shown in Figure 2-32. The holes are, however, angled outward approximately 5–10 degrees. This places the zone of grouted material outside the excavation limits when shaft excavation is resumed. The shaft can also be overexcavated at the grouting location to allow the holes to be drilled vertically. The grouting operation first consists of completely mucking, removing loss material from the shaft than blowing or cleaning the shaft bottom. Next a 0.3 m (1 ft) thick 20 Mpa (3,000 psi) plain concrete leveling/working pad is placed over the entire shaft bottom. The grout cover or ring is then drilled and grouted (see Fig. 2-34). After the grout has had time to set the shaft excavation is resumed (Goff et al. 1985). The complete in-shaft grouting operation as described above becomes part of the critical path of the shaft excavation and lining installation activities. It will, therefore, have a day-for-day impact on the overall shaft development schedule.

In-shaft grouting is also performed for inclined shafts, which are shafts excavated at an angle with the horizontal. Most inclines are excavated at angles of 45 degrees or less. Excavation can be advanced in the upward or downward direction from the horizontal. When these shafts are excavated downward they are sometimes called declines. The grout holes are drilled to approximately 10 m (30 ft) in depth ahead of the working face of the incline shaft. Each of the grout rings or grout covers are overlapped by approximately 1 m (3 ft) (see Fig. 2-35). The hole layout and grouting sequence are basically the same as those used for pregrouting, as shown in Fig. 2-32. The holes are, however, angled outward approximately 5–10 degrees. This places the zone of grouted material outside the excavation limits when the incline shaft excavation is resumed.

GROUTING METHODS

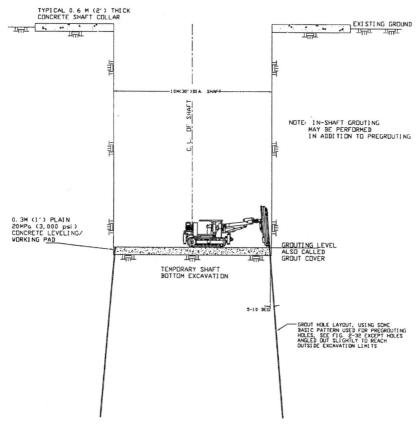

FIG. 2-34. Section Showing Typical In-Shaft Grouting Arrangement

Isolated or unexpected water inflows can be grouted on an as-needed basis, as they are encountered during shaft and incline shaft excavation. When a water-bearing zone is encountered in a shaft, horizontal or angled holes are drilled radially and grouted as needed to control the inflow of water. In the case of incline shafts, the holes are drilled in the water-bearing zone at an orientation to maximize the effect of the grouting. Grout nipples with valves are installed in the shaft walls directly into the water-bearing zones. Packers at the hole collar can also be used (see Fig. 2-36). The hole depths can vary depending on the geology and structure of the water bearing zone. However, a 3 m (10 ft) hole depth is a good rule of thumb. As with preplanned in-shaft grouting, any isolated or unexpected grouting operation performed in the shaft will have a day-for-day impact on the overall shaft development schedule.

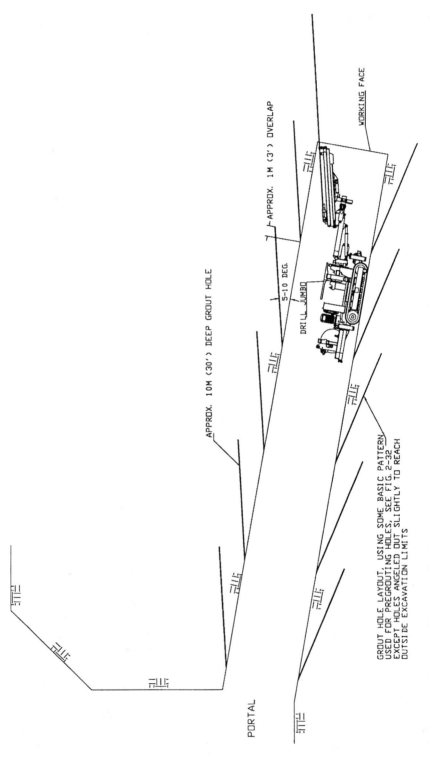

FIG. 2-35. Section of Typical In-Shaft Grouting Arrangement Used in Inclined Shaft

GROUTING METHODS

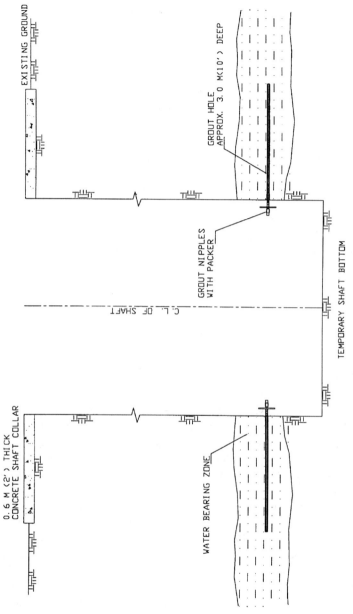

FIG. 2-36. Section of Typical In-Shaft Grouting Using Horizontal Holes with Grout Nipples and/or Grout Packers

2.5 PROBING AND GROUTING AHEAD OF TUNNEL AND CHAMBER EXCAVATIONS

Probing ahead of the working face during tunnel and chamber excavations is done to explore for groundwater which may be present in unacceptably large quantities and or elevated pressures. The probe hole is also used to detect changes in the geologic structure and type of materials which may exist immediately in front of the advancing excavation. Probe holes are also used to check for elevated levels of explosive and poisonous gases. Checking for gas levels is necessary, and should be required in the specifications on projects where the geologic setting is conducive to the formation of gas or when the history of other excavations in the area suggests the possibility of encountering gas. Also, the industrial development of the surrounding area should not be overlooked as a possible source of gas or other hazardous materials. For example, current and abandoned gasoline filling stations, underground storage tanks, chemical settlement ponds, and pipelines within the project area should be considered potential hazards during the design and construction stages of a project.

Grouting ahead of the working face is performed to stop or substantially reduce the flow of groundwater into the excavation. Grouting ahead is also used to stabilize or modify the in situ geologic material to be excavated. It can also be used to control the flow of gases into the excavation.

2.5.1 Probe Holes

Probe holes are also called feeler holes. They are drilled to some distance, usually 10–40 m (30–130 ft), ahead of the excavation working face. Probe holes are used for excavations performed by the drill-and-blast, roadheader, TBM, shield, and other types of mechanical excavation methods. Except for TBM excavations, most probe holes are drilled directly into the working face. The probe hole does not have to be drilled precisely in the center of the working face when equipment or other logistical requirements prevent it. However, an effort should be made to keep the probe hole within 1–2 m (3–6 ft) of the center of the working face. When the size of the working face is large or geologic conditions warrant, the use of several probe holes may be justified. Probe holes are normally drilled parallel to the alignment of the advancing excavation. If, however, a geologic structure or condition which might require grouting is thought to exist, probe holes are drilled at orientations designed to intersect and most effectively grout the suspected zones.

When probe holes are used in conjunction with TBM excavations the current practice is for the drill(s) used for probe and grout hole drilling to be located on the TBM at some distance, 3–5 m (10–16 ft), back from the working face. The distance depends on the diameter of the tunnel and the TBM manufacturer's machine layout. Using this method, the probe hole is drilled at a slight angle outward from the tunnel alignment. Figs. 2-37 and 2-38 show a general equipment layout and probe/grout hole orientation. This arrangement allows the probe hole to be drilled without necessitating drill rods going through an opening in the TBM cutterhead. An entire probe hole can usually not be completed while the TBM is mining. In an effort to minimize the loss of TBM production, the majority of a probe hole can be drilled during the TBM maintenance shift. To further minimize any loss in TBM production, the hole can be started during the last one or two TBM advances before maintenance begins. Likewise, the hole can be completed during the first or second TBM advancement following the completion of maintenance. The capability to drill while TBM mining is made possible by using drills mounted on slides that allow the drill to remain stationary while the TBM is advancing. The travel on the drill slide is usually limited to 3.0–3.5 m (10–12 ft). If this type of equipment arrangement were not used and probe holes were to be drilled directly into the working face, through an opening in the TBM cutterhead, the TBM advance would have to be stopped during the entire probe hole drilling operation. It can easily be seen that a shutdown of TBM mining is particularly costly to production if a total grout cover, consisting of say 14 holes, were drilled and grouted through the cutterhead. It is therefore desirable to use a drilling and grouting equipment arrangement which maximizes TBM mining time.

In their manufacturer's literature for the Jarva MK-27 TBM, the Robbins Company states that probe/grout holes can be started or collared at as small an angle as 7 degrees to the rock (see Figs. 2-37 and 2-38). For the Hallandsasen Railway Tunnel Project in Sweden, a MK-27 TBM was equipped with four hydraulic rock drills, one for each of the four quadrants. The drills were carried on rota-extension booms and capable of drilling holes 40 m (130 ft) long with the hole bottom at a maximum of 5 m (16 ft) outside the further TBM bore. The Hallandsasen Tunnel was to be excavated to 9.1 m (28 ft) in diameter. Fig. 2-39 is an artist's rendering of an MK-27 TBM showing two of the four rock drills. On the Boston Harbor Inter-Island Tunnel, probe hole drilling was initiated 4 m (12 ft) back from the TBM tunnel face at an angle of 15 degrees to the rock. The average hole depth was 19.5 m (65 ft) using a 56 mm (2-1/4 in.) diameter bit.

2.5.2 Grouting ahead of Excavations

Grouting ahead of the working face is performed to control groundwater inflows and to stabilize or modify the ground. It can also be used to control gas and other hazardous material from entering the excavation. The need to perform grouting in advance of the excavation is based on the results of probe hole drilling or conditions encountered during excavation. The need

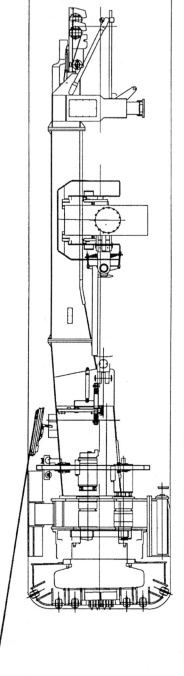

FIG. 2-37. General Equipment Layout of Probe/Grout Hole Drill on TBM

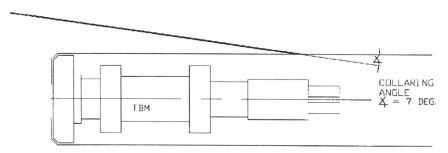

FIG. 2-38. Section Showing Probe/Grout Hole Collaring Angle

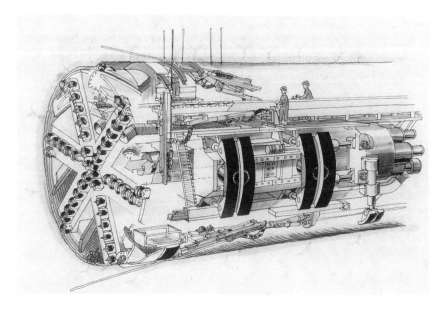

FIG. 2-39. MK-27 TBM Showing Two of Four Drills (Courtesy of Robbins Company 1994)

to grout may also be known before any excavation begins from data collected during the geotechnical site investigation.

When the need to grout is known in advance, the possibility of grouting from the surface should be thoroughly evaluated. However, grouting from the surface may be impractical due to the depth of the grout zone, rough terrain, inaccessibility, or surface obstructions, such as industrial and urban development. Because grouting from the surface is an operation independent from the excavation, it can be performed in advance of the underground excavation reaching the area which requires grout. Therefore, with proper planning, it can be scheduled to be performed so as not to delay the excavation.

When grouting is performed ahead of the working face, from within the tunnel or chamber heading excavation, it has a negative impact on the excavation schedule. The negative impact on TBM-excavated tunnels can be somewhat minimized by using drilling equipment and methods similar to these designed for the Hallandsasen Railway Tunnel and the Boston Harbor Inter-Island Tunnel. On the Hallansaen project the 9.1 m (28 ft) diameter tunnel was to be grouted ahead of the working face using 14 evenly spaced grout holes. During TBM excavation of the Boston project, a total of 48 probe holes were drilled over a tunnel distance of 370 m (11,308 ft). However, of the 48 holes drilled, only five required grouting. The project specifications required only probe holes with groundwater inflows of 188 L/min (50 gal./min) or higher to be grouted. An average drilling time of 12 h for probe holes up to 40 m (130 ft) in depth was reported for this project.

CHAPTER 3

GEOTECHNICAL CONSIDERATIONS IN GROUTING PROGRAM PLANNING[a]

As discussed in Chapter 2, underground grouting operations are either geotechnically or structurally oriented. Geotechnical grouting entails jet, compaction, permeation, hydrofracture, consolidation, and curtain grouting. As the term implies, geotechnical grouting involves filling the voids of the geological material, that is, the soil or rock, surrounding the structure. This serves to decrease flow of water into and out of a structure, to increase the strength of soil, and to enhance the stability of the ground. Structural grouting, on the other hand, deals with filling the voids between a structure and the ground. It is also used to fill in defects or voids in structural concrete elements that result from poor placement procedure, as well as the fluid behavior of concrete before it sets. Structural grouting methods include contact, embedment, and prestressing grouting.

Most often, geotechnical grouting is used to control the inflow of groundwater, which, if not controlled, can have a profound affect on the stability of the opening during excavation, can increase construction costs, and can adversely affect the performance of the final product. When grouting is employed to control water, the stability and strength of the in situ foundation material are also augmented. In some cases, therefore, one grouting operation can address these concerns simultaneously.

There are situations when groundwater is not a problem yet grouting is still required. Examples would be when weak soils require strengthening or cohesionless soils above the water table require stabilization. From a structural standpoint, it can be stated that contact grouting is required whenever tunnel and chamber designs include liners. Embedment and prestressing grouting, however, are called for only when certain structural components are included in the design or high water pressure is expected. Chapter 2 gives a more detailed description of the various grouting methods and their uses.

[a]This chapter was contributed by Thomas M. Saczynski.

3.1 BASIC CONSIDERATIONS

Preparing a geotechnical grouting program requires consideration during the early design phase of a project. It is worth repeating here that the grouting program must remain flexible to allow modifications during the construction phase to suit actual site conditions. The designer must develop criteria for the allowable amount of groundwater infiltration and devise a grouting design. The designer must also assess the stability of the structure and determine if the stability must be augmented by grouting and devise a grouting design that will function according to the developed criteria. In the construction phase, the contractor must determine the means and methods to meet the design requirements.

The grouting plan must be based on the nature of the ground surrounding the structure. Ground characterization is determined from a geotechnical site investigation. The ensuing analysis and conclusions are found in the geotechnical report or geotechnical design summary report (GDSR) for each specific project.

3.2 GEOTECHNICAL INVESTIGATION AND GEOTECHNICAL DESIGN SUMMARY

The geotechnical investigation is performed to secure sufficient design information about the ground that the structure is located within. Specific information must be obtained to design an excavation that will be stable both during and after construction. This includes the determination of the loads that will be imposed on the structure by soil, rock, and groundwater. The information is used for the design of the excavation support, the selection of construction methods, and the design of the final structure. The information also provides the basis for the formulation of a grouting plan.

The results of geotechnical investigations are contained in the GDSR, which should be made a part of the contract documents for underground projects. The GDSR contains a detailed description of the ground conditions found during the investigation. Although the contents of GDSRs vary considerably in scope, format, and detail, there are certain attributes that are typical to most. The parameters required for determining the necessity of grouting, that is, permeability, strength, and stability, can be developed from these attributes.

A GDSR generally contains a discussion of the site geology, a treatment of previous work in the area, a qualitative overview of the anticipated behavior of the soil or rock as it is excavated, construction considerations, a graphic presentation of the project alignment with respect to the various geologic layers, boring logs, design assumptions, design parameters, and test data. While all of these are useful for the planning of a grouting operation, the test data from permeability testing is the most valuable.

The capacity of a geologic medium to transmit water is designated as the permeability of that medium. Establishing the permeability of the geologic material surrounding the structure is essential for approximating the amount of water that will flow into the structure. Should calculations demonstrate that unacceptable flows will be encountered either during or after construction, the designer may specify a grouting program in the contract documents. A grouting program may, on the other hand, be left to the contractor to develop to control water inflow during construction. In either case the permeability testing and results should be referred to in the narrative portion of the GDSR with particular attention devoted to the potential construction ramifications envisioned with and without a grouting program.

The permeability of soils depends on the gradation, density, degree of saturation, and stratification. Mineralogy and structure are also important in clays. The permeability of a rock mass centers on the nature of the discontinuities present, that is, the joints, fractures, shears, faults, and bedding planes. Intact rock permeability, a function of the porosity, is generally much less permeable than the rock mass.

The designer must keep in mind that geologic materials are most often heterogeneous and anisotropic, that is, they vary substantially in composition and their engineering properties vary directionally. An example of a heterogeneous condition is a soil profile with clay underlying sand with lenses of silt interspersed throughout. An example of an anisotropic condition is a soil that has a certain permeability for vertical flow, but exhibits a different permeability for horizontal flow.

Testing soil for permeability is accomplished through constant-head tests, falling-head tests, and rising-head tests. Constant-head tests are used for estimating the permeability of coarse-grained soils of high permeability, such as clean sands and gravel. Falling-head tests are used for fine-grained soils with low permeability, such as silty or clayey fine sand, silts, and clays. These tests may be performed in situ or in the laboratory. For procedural and computational details see NAVFAC *Design Manual DM-7* (1971) or Hoek and Bray (1977).

Laboratory testing, however, poses some limitations. Generally speaking, it can not by itself lead to conclusions regarding the behavior of the entire soil mass around the structure. Not only is the sample disturbed when it is removed, but the sample is quite unlikely to be representative of the soil mass as a whole. Therefore, in situ tests are indispensable for developing an understanding of the potential soil mass behavior during construction and over the life of the structure.

Rock testing should be performed by pressure testing the in situ rock mass. Laboratory testing will only result in the permeability of the sample, which is a function of the porosity. This is of limited use as the discontinuities of the rock mass, rather the than intact rock, transport water.

Permeability testing for underground work should be conducted from within exploratory boreholes that are used to determine other geotechnical parameters. In selecting a test interval, the designer should focus upon an envelope extending a particular distance above and below the structure. For tunnels, this is about one tunnel diameter above the tunnel crown to one diameter below the structure.

Soil sampling or rock coring may indicate zones that typically characterize extremes of very high or low permeability outside of the one-diameter envelope recommended above. This could be a pocket of gravel or a shear zone of highly fractured rock, which typically show high permeability, or a clay lense or massive, unfractured rock, which shows low permeability. Even though outside the envelope, these extreme zones could affect the grout design depending on the orientation and extent with respect to the structure alignment. The designer must therefore determine from the manner in which these extremes fit into the graphic presentation of the GDSR as to whether or not additional pressure testing is required beyond the recommended one-diameter envelope.

For in situ pressure testing, an increment of the borehole is isolated with a two-packer or double-packer arrangement, as described in section 4.8 of chapter 4. Water is injected at a specified pressure into a borehole increment (Fig. 3-1). The amount of water lost to the geological material at a particular pressure over a time interval, called "take," is recorded. This will lead to the calculation of a permeability coefficient k expressed in terms of cm/s (in./sec). Again, for procedural and computational details see NAVFAC *Design Manual DM-7* (1971) or Hoek and Bray (1977).

In Europe, pressure testing incorporates a unit called a "lugeon," which is defined as a flow of 1 L of water per minute per meter of length of borehole length at a pressure of 10 kg/cm^2 (142 psi). Houlsby (1986) offers a sophisticated scheme for determining the need for grouting beneath dams. Although the refinement offered by Houlsby for dam grouting has no parallel for underground structures, Doig (1985) suggests the following grouting needs for corresponding Lugeon values:

- 1–3 Lugeons: no grouting
- 3–10 Lugeons: chemical grouting only
- 10 or more Lugeons: cement grout acceptable

Heuer (1995) also discusses lugeon testing for underground work. The designer must exercise judgment in choosing the method for determining the need for a grouting program. Although the simplicity of Doig's recommendations is most attractive when estimating the amount of "take" for a particular grout hole, a more rigorous assessment may be warranted, as discussed in the following section.

GEOTECHNICAL CONSIDERATIONS IN GROUTING PROGRAM PLANNING

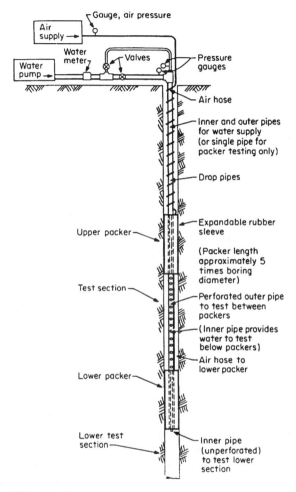

FIG. 3-1. Apparatus for Pressure Testing a Borehole (Hunt 1984; Courtesy of McGraw-Hill)

Although permeability testing is the most important element of the GDSR with respect to grouting, it actually tests only a small percentage of the project in terms of the gross volume. To extrapolate the character of the entire project based solely on the permeability presented is not prudent. Other portions of the GDSR must be considered to develop a more accurate frame of reference for the project as a whole. The rest of the GDSR usually contains information vital to achieving a good overall understanding of the geological setting of the project.

The discussion of the geology presents such information as the character of the overburden soil as well as that of the bedrock. Quantitative approximations of the more pertinent aspects of the project can be developed as well as the qualitative behavior of the material. For example, knowing that a particular soil layer is a lacustrine deposit consisting of silty clay suggests that this layer will be relatively impervious. A rock formation that is described as having many joint sets with closely spaced joints suggests that it may be relatively permeable.

The graphic portrayal of the project alignment within the various geologic horizons demonstrates the changing character of the ground to be excavated (Fig. 3-2). It may be helpful in predicting where grouting may be required to prevent a sudden inflow of water, as in faulted zones, or zones where grouting may cease to be effective, as in clay. Such indications are indispensable when planning a grouting program.

The borehole data presented in boring logs is most useful. If there has been permeability testing performed on the boring, the permeability coefficient will be presented in the description at the appropriate elevation. Information is also shown that contains a physical description; the soil classification; density or consistency; and the blow count from a standard penetration test (Fig. 3-3). When the borehole passes through rock the information will contain the rock quality designation for the interval, as well as a physical and geologic description of the rock (Fig. 3-4).

3.3 GROUTING TO LIMIT GROUNDWATER INFILTRATION

The designer must determine how much groundwater infiltration is allowable. Typical values for allowable infiltration rates are approximately 0.8 L/m^2/day (0.02 gal./ft^2/day) for transit projects and 15 L/m^2/day (0.4 gal./ft^2/day) for wastewater conveyance projects (O'Rourke 1984). The actual determination of the allowable infiltration rate is specific to the individual project. These needs become apparent as the project evolves from the initial concept through final design.

The actual rate of infiltration into an underground structure is a function of the location of the water table with respect to the structure, the geology surrounding the structure, the operating pressure of the structure, and the integrity of the lining of the structure. The water table, as well as the character of the soil and rock, is determined during the geotechnical investigation. The lining and operating pressure of the structure are structural considerations rather than geotechnical. The lining may be assumed to be impervious for the purpose of geotechnical planning.

The permeability of a geologic medium is determined so that the quantity of flow into a structure can be computed. A good approximation developed by Goodman et al. (1965) for the flow into a circular tunnel is given by the formula

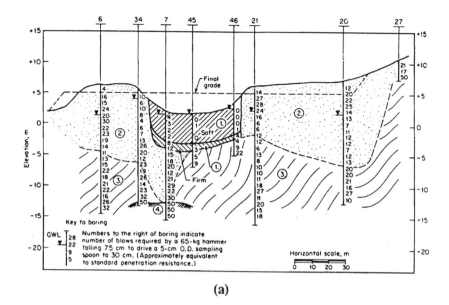

(a)

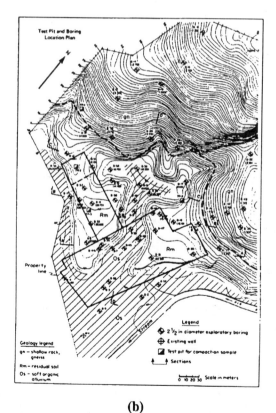

(b)

FIG. 3-2. Geologic Interpretation from Borehole Data (Hunt 1984; Courtesy of McGraw-Hill)

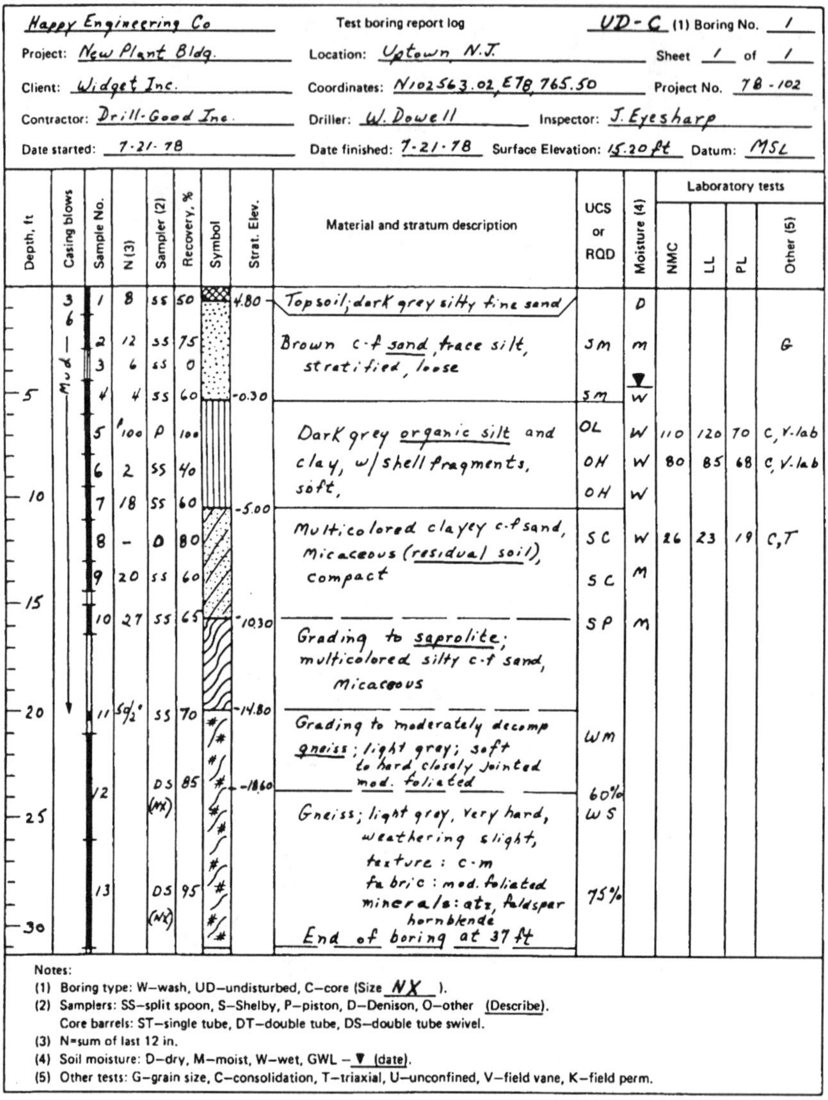

Happy Engineering Co — Test boring report log — UD-C (1) Boring No. 1

Project: New Plant Bldg. — Location: Uptown N.J. — Sheet 1 of 1

Client: Widget Inc. — Coordinates: N102563.02, E78765.50 — Project No. 78-102

Contractor: Drill-Good Inc. — Driller: W. Dowell — Inspector: J. Eyesharp

Date started: 7-21-78 — Date finished: 7-21-78 — Surface Elevation: 15.20 ft — Datum: MSL

Depth, ft	Casing blows	Sample No.	N (3)	Sampler (2)	Recovery, %	Strat. Elev.	Material and stratum description	UCS or RQD	Moisture (4)	NMC	LL	PL	Other (5)
	3, 6	1	8	SS	50	4.80	Topsoil; dark grey silty fine sand		D				
	Mud	2	12	SS	75		Brown c-f sand, trace silt, stratified, loose	SM	M				G
		3	6	SS	0								
5		4	4	SS	60	-0.30		SM	▼ W				
		5	100	P	100		Dark grey organic silt and clay, w/shell fragments, soft,	OL	W	110	120	70	C, V-lab
		6	2	SS	40			OH	W	80	85	68	C, V-lab
10		7	18	SS	60	-5.00		OH	W				
		8	-	O	80		Multicolored clayey c-f sand, Micaceous (residual soil), compact	SC	W	26	23	19	C, T
		9	20	SS	60			SC	M				
15		10	27	SS	65	-10.30	Grading to saprolite; multicolored silty c-f sand, Micaceous	SP	M				
20		11	50/2"	SS	70	-14.80	Grading to moderately decomp gneiss; light grey; soft to hard, closely jointed mod. foliated	WM					
		12		DS (NX)	85	-18.60	Gneiss; light grey, very hard, weathering slight, texture: c-m fabric: mod. foliated minerals: qtz, feldspar hornblende	60% WS					
25													
		13		DS (NX)	95			75%					
30							End of boring at 37 ft						

Notes:

(1) Boring type: W–wash, UD–undisturbed, C–core (Size NX).

(2) Samplers: SS–split spoon, S–Shelby, P–piston, D–Denison, O–other (Describe). Core barrels: ST–single tube, DT–double tube, DS–double tube swivel.

(3) N=sum of last 12 in.

(4) Soil moisture: D–dry, M–moist, W–wet, GWL –▼ (date).

(5) Other tests: G–grain size, C–consolidation, T–triaxial, U–unconfined, V–field vane, K–field perm.

FIG. 3-3. Typical Borehole Data in Soil (Hunt 1984; Courtesy of McGraw-Hill)

$$Q_0 = [2\pi k H_0] \div [2.3\log(2H_0/r)] \qquad (3.1)$$

in which Q_0 = flow in terms of volume per unit of time per unit length of tunnel; k = the permeability constant in terms of length per unit of time; H_0 = distance from the water table to the invert of the tunnel; and r = radius of

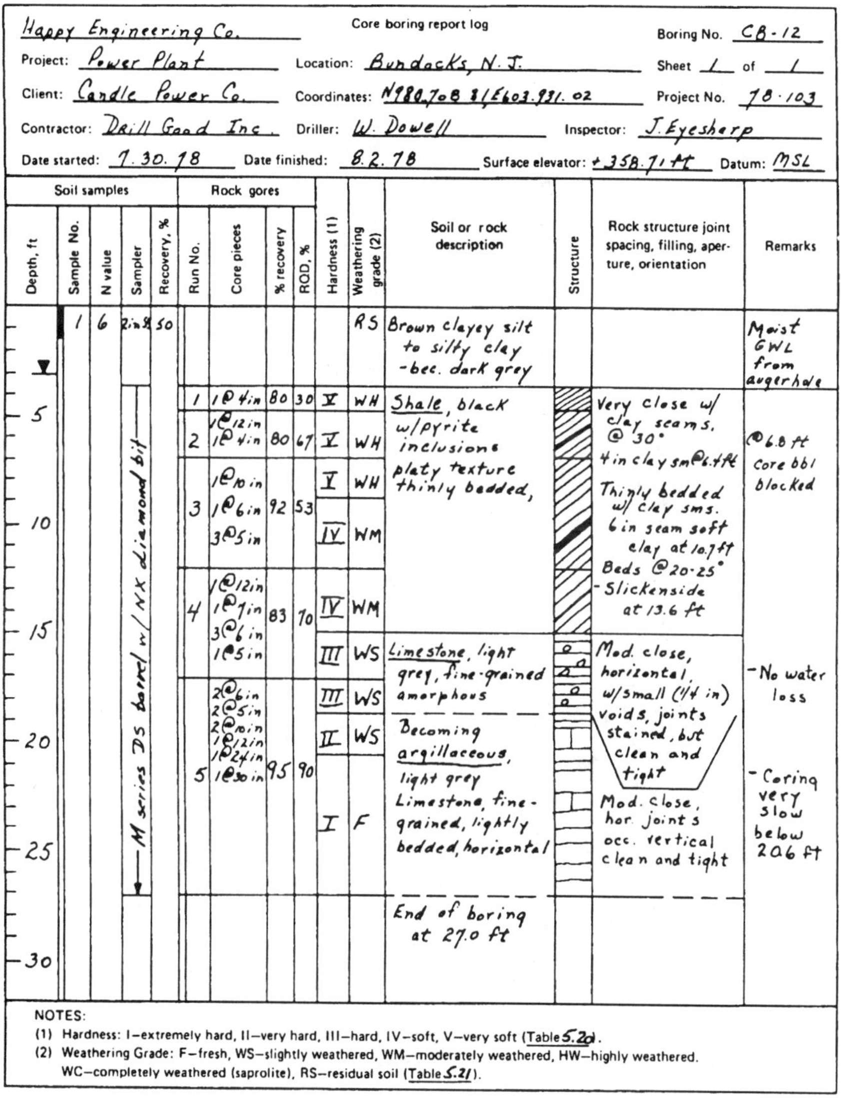

GEOTECHNICAL CONSIDERATIONS IN GROUTING PROGRAM PLANNING

```
Happy Engineering Co.          Core boring report log          Boring No. CB-12
Project: Power Plant           Location: Bundocks, N.J.        Sheet 1 of 1
Client: Candle Power Co.       Coordinates: N980.708 E/E603.931.02   Project No. 78-103
Contractor: Drill Good Inc.    Driller: W. Dowell              Inspector: J. Eyesharp
Date started: 7.30.78          Date finished: 8.2.78           Surface elevator: +358.71 ft   Datum: MSL
```

Depth, ft	Soil samples				Rock cores				Hardness (1)	Weathering grade (2)	Soil or rock description	Structure	Rock structure joint spacing, filling, aperture, orientation	Remarks
	Sample No.	N value	Sampler	Recovery, %	Run No.	Core pieces	% recovery	ROD, %						
	1	6	2 in	50						RS	Brown clayey silt to silty clay - bec. dark grey			Moist GWL from augerhole
5					1	1@4in	80	30	V	WH	Shale, black w/pyrite inclusions platy texture thinly bedded,		Very close w/ clay seams @ 30° 4 in clay sm @ 6.4 ft	@ 6.8 ft core bbl blocked
					2	1@12in 1@4in	80	67	V	WH				
10					3	1@10in 1@6in 3@5in	92	53	V IV	WH WM			Thinly bedded w/ clay sms. 6 in seam soft clay at 10.7 ft Beds @ 20-25° - Slickenside at 13.6 ft	
15					4	1@12in 1@7in 3@6in 1@5in	83	70	IV III	WM WS	Limestone, light grey, fine-grained amorphous		Mod. close, horizontal w/small (1/4 in) voids, joints stained, but clean and tight	- No water loss
20					5	2@6in 2@5in 2@10in 1@12in 1@24in 1@30in	95	90	III II I	WS WS F	Becoming argillaceous, light grey Limestone, fine-grained, lightly bedded, horizontal		Mod. close, hor. joints occ. vertical clean and tight	- Coring very slow below 20.6 ft
25														
30											End of boring at 27.0 ft			

NOTES:
(1) Hardness: I—extremely hard, II—very hard, III—hard, IV—soft, V—very soft (Table 5.2a).
(2) Weathering Grade: F—fresh, WS—slightly weathered, WM—moderately weathered, HW—highly weathered. WC—completely weathered (saprolite), RS—residual soil (Table 5.2f).

FIG. 3-4. Typical Borehole Data in Rock (Hunt 1984; Courtesy of McGraw-Hill)

the tunnel. Permeability constants can be estimated from published tables and graphs, as well as determined from laboratory testing, or in situ testing. Typical permeability coefficients for soil and rock are shown in Table 3-1 and for some natural soil formations in Table 3-2.

Even though in situ testing is the preferred method to determine perme-

TABLE 3-1. Typical Permeability Coefficients for Rock and Soil Formations (Hoek and Bray 1977)

Description (1)	k (cm/s) (2)	Intact Rock (3)	Porosity (%) (4)	Fractured rock (5)	Soil (6)
Practically impermeable	10^{-10} 10^{-9} 10^{-8} 10^{-7}	Massive low-porosity rocks	0.1–0.5 0.5–5.0		Homogeneous clay below zone of weathering
Low discharge, poor drainage	10^{-6} 10^{-5} 10^{-4} 10^{-3}	Weathered granite Schist Sandstone	5.0–30	Clay-filled joints	Very fine sands, organic and inorganic silts, mixtures of sand and clay, glacial till, stratified clay deposits.
High discharge, free draining	10^{-2} 10^{-1} 10^{0} 10^{1} 10^{2}			Jointed rock Open-jointed rock Heavily-fractured rock	Clean sand, clean sand and gravel mixtures. Clean gravel

TABLE 3-2. Permeability Coefficients for Some Natural Soil Formations (Terzaghi and Peck 1967)

FORMATION (1)	VALUE of k, (cm/s) (2)
River deposits	
Rhone at Genissiat	Up to 0.40
Small streams, eastern Alps	0.02–0.16
Missouri	0.02–0.20
Mississippi	0.02–0.12
Glacial deposits	
Outwash plains	0.05–2.00
Esker, Westfield, Mass.	0.01–0.13
Delta, Chicopee, Mass.	0.0001–0.015
Till	$<.0001$
Wind deposits	
Dune sand	0.1–0.3
Loess	$0.0001\pm$
Loess loam	$0.0001\pm$
Lacustrine and marine offshore deposits	
Very fine uniform sand, (C_u = 5 to 2)	0.0001–0.0064
Bull's liver, Sixth Ave., N.Y., C_u = 5 to 2	0.0001–0.0050
Bull's Liver, Brooklyn, C_u + 5	0.00001–0.0001
Clay	$<.0000001$

ability, the designer must again be aware that the geologic material in question will most likely be heterogeneous and anisotropic. In practical terms, this means that the permeability coefficient determined can vary. The author has experienced that the permeability actually encountered can deviate from that predicted based on pressure testing by a half of an order of magnitude.

3.4 GROUTING TO INCREASE STABILITY

Certain ground conditions may require grouting to stabilize the earth mass around an excavation. For example, when the alignment of a tunnel passes through cohesionless soil, some measures are needed to provide stability of the working face. An excavation in poor quality rock, which can behave like a cohesionless soil, may also require additional measures to ensure stability.

When excavating above the water table in cohesionless soil, a vertical working face can not be maintained and the soil will move into the excavation opening. If nothing is done to prevent the soil from moving, it will continue to do so until the slope within the excavation becomes stable. This occurs when the slope reaches its natural angle of repose, which is the angle of internal friction of the soil. The ground loss at the working face allows a void to form that stopes, or propagates, to the surface with the net result of subsidence of the ground surface. Depending on the speed with which the working face collapses, cohesionless soil above the water table is termed either "running ground," "fast ravelling ground," or "slowly ravelling ground" (Proctor and White 1977).

When excavating below the water table in cohesionless soil, the working face is likely to be classified as "flowing ground," where water will flow into the working face carrying soil with it. Flowing ground is difficult to stop once it has been encountered in the working face. (Proctor and White 1977).

The presence of fine-grained soils, that is, silts and clays, lends stability to the working face of an underground opening. The fines tend to impart an "apparent cohesion" within the soil mass thereby increasing the "stand up time" of the face (Proctor and White 1977). Whereas a clean sand below the water table would result in flowing ground, a clayey sand would result in cohesive running, fast ravelling, slow ravelling, or firm ground, depending on the amount of clay present. Tables 3-3, 3-4, and 3-5 summarize the ground classifications for soft-ground tunneling and the corresponding behavior of the working face.

Where running or flowing ground is anticipated, a breasting system may be incorporated. This entails some kind of shoring of the vertical working face (Proctor and White 1977). Regardless of the sophistication of the breasting system some, loss of ground into the opening will occur. This will ultimately result in surface subsidence (Peck 1969). When the amount of predicted subsidence is unacceptable to the designer, special construction methods or ground modification become necessary. Construction methods might include excavating under compressed air, use of an earth pressure balance TBM, or using a slurry shield TBM. Ground modification may include grouting, dewatering, ground freezing, or a combination of these methods.

Grouting is an alternative to soil freezing and compressed air for stabilizing the working face. A widely used method of soil stabilization is permeation grouting. In this method, grout is injected into the soil voids and hardens, thereby cementing the soil particles and imparting a cohesive character to the soil. The net result is a behavioral change from running or flowing ground into firm ground.

TABLE 3-3. Working Face in Silty Sand (Proctor and White 1977)

Designation (1)	Effective grain size D_{10} (mm) (2)	Degree of compactness (3)	Probable behavior in tunnel		
			Above water table (4)	Below water table	
				Free air (5)	Compressed air (6)
Fine silty sand (uniformity coefficient < 3)	0.05	Dense, $N > 30$ Loose, $N < 10$	Rapidly ravelling Cohesive running	Flowing ground	Rapidly ravelling Cohesive running ground
Silty sand (uniformity coefficient > 6)	0.05	Dense, $N > 30$ Loose, $N < 10$	Slowly ravelling Cohesive running	Flowing ground to cohesive-running ground	Rapidly ravelling Cohesive running ground

TABLE 3-4. Working Face in Sand or Sandy Gravel with Binder (Proctor and White 1977)

Designation (1)	Degree of compactness (2)	Probable behavior in tunnel		
		Above water table (3)	Below water table	
			Free air (4)	Compressed air (5)
Fine sand with clay binder	Dense, $N > 30$ Loose, $N < 10$	Firm or slowly ravelling Rapidly ravelling	Slowly ravelling Flowing	Firm or slowly ravelling Rapidly ravelling
Sand or sandy gravel with clay binder	Dense, $N > 30$ Loose, $N < 10$	Rapidly ravelling Rapidly ravelling	Firm or slowly ravelling Rapidly ravelling or flowing	Firm Rapidly ravelling

TABLE 3-5. Working Face in Clean Gravel and Sand (Proctor and White 1977)

Designation (1)	Grain size D10 (mm) (2)	Probable behavior in tunnel		
		Above water table (3)	Below water table	
			Free air (4)	Compressed air (5)
Gravel	> 2.0	Running ground	Flowing with extremely heavy inflow of water	Running ground, excessive loss of air
Sand gravel	> 0.2	Running ground	Flowing with extremely heavy inflow of water	Running or cohesive-running ground with heavy loss of air
Coarse-to-medium sand	2.0–0.2	Running ground	Flowing with extremely heavy inflow of water	Running or cohesive-running ground with heavy loss of air
Very fine sand	0.1–0.05	Fast ravelling (Dense, $N > 30$) Cohesive running (Loose, $N < 10$)	Flowing ground	Rapidly ravelling ground with moderate loss of air

Of utmost importance is the even distribution of the grout throughout the soil mass treated. The mere injection of grout under pressure is likely to form isolated bulbs of grout, thereby leaving large pockets of soil untreated. When the working face meets such untreated soil, running or flowing ground is a likely consequence. To avoid this problem, field testing of grout mixes and procedures should be performed to optimize grout travel and to enhance the even distribution of grout.

3.5 GROUTING TO INCREASE STRENGTH

The conditions that require a designer to examine the stability of the working face of an underground opening may also require a plan to strengthen the soil. Excavating in soils with little stand-up time may require soil modification so that an arch is formed to carry the entire overburden load as an excavation proceeds. The arch must carry this load until the lining is erected. In the case of a circular tunnel, the designer can model the excavation as a circular hole cut within a stress field, compute the resulting stress concentrations, and compare these stresses with the shear strength of the modified soil. The actual shear strength used must be determined from samples taken from the modified soil.

For planning purposes, neat cement grout can develop 7–10 MPa (1,000–1,500 psi) of unconfined compressive strength. Chemical grouts can develop 35–350 kPa (5–50 psi) of "useable" strength, with a factor of safety of 2.0 (Karol 1983). More recent experience, however, suggests that strengths approaching 3.5 Mpa (500 psi) can be achieved (see Chapter 10).

As with stability, the even distribution of the grout throughout the soil mass treated is critical. Since areas of untreated soil will run into the tunnel, the formation of a void tends to defeat the desired arching effects created by the cemented soil.

3.6 GROUTABILITY RATIO

The designer may determine that the in situ conditions are not acceptable from the standpoint of permeability, stability, or strength, and that permeation grouting appears to be an attractive remedial measure. He must then determine if grouting is indeed possible. The groutability ratio GR is a useful parameter for checking the applicability of portland cement grout for use in sand and is given by the formula (Mitchell 1970)

$$GR = D_{15}/D_{95} \qquad (3.2)$$

where GR = the groutability ratio for the soil to be grouted; D_{15} = the particle diameter of the soil to be grouted 15% of which is finer by weight; and D_{95} = the particle diameter of the grout 95% of which is finer by weight.

Weaver (1991) summarizes the possibility of grouting a soil for GR ranges as $GR > 24$, usually; $GR < 19$, not likely; and $GR < 11$, not possible. Filling in the gaps leads one to surmise that for $19 < GR < 24$, grouting is difficult, but possible; and $11 < GR < 19$, possible, but not likely.

For portland cement grouting of rock, the groutability ratio is given by the formula

$$GR = \text{width of fissure}/ (D_{95}) \text{ grout.} \qquad (3.3)$$

The possibility of grouting rock is $GR > 5$, grouting consistently possible; $GR < 2$, grouting not possible.

The groutability of soil or rock using chemical grouts is given in terms of the permeability constant, k in cm/s (Karol 1983): $k < 10^{-6}$, ungroutable; $10^{-6} < k < 10^{-5}$, difficult grouting with grout viscosity less than 5 centepoises, impossible with viscosity greater than 5 centepoises; $10^{-5} < k < 10^{-3}$, difficult grouting with viscosity greater than 10 centepoises; $10^{-3} < k < 10^{-1}$, groutable with all commonly used chemical grouts; and $k > 10^{-1}$, use suspended solids grout or chemical grout with filler. See Chapter 10 for further discussion of chemical grout.

3.7 OTHER GEOTECHNICAL CONSIDERATIONS

There are other qualitative considerations that should be understood when planning for a grouting operation. As described extensively in his work, Weaver (1991) recounts such characteristics as hydraulic routing, surface roughness, tortuosity, fracture aperture, porosity, and permeability. These are attributes of a rock mass that will affect the grouting operation, but are difficult to assess by way of computation.

The hydraulic route, that is, the course water flows through in rock, has some affect on groutability. This is due to the nature of the discontinuities carrying the water (Ewart 1958). Although we may model a joint or fracture as a continuous and regular plane to simplify analysis, this is seldom the case. In reality, a discontinuity ends at some point, the walls are rarely planes, and the wall surfaces are usually irregular. The path of water follows the path of least resistance. When restricted, it flows into an intersecting discontinuity. Also, water flowing through irregular openings erodes the opening walls, thereby increasing their width. This is why the development of expanded openings tends to occur along the intersections of planar features and in softer rock.

This phenomenon is an example of the heterogeneous nature of rock masses. Because of heterogeneity, it is quite difficult to predict where a grout hole will intersect a water-bearing plane. As Weaver stated, "any intersection of such a feature by a grout hole would be extremely fortuitous." This is the reason for the grout hole patterns elaborated upon in Chapter 2.

The roughness of the hydraulic route is also a consideration. In geologic settings where the surfaces of the discontinuities are rough, a head loss due to friction results, which in turn causes a pressure drop. Grout particles then settle out in low spots from the drop in pressure. The observed outcome is a diminished amount of grout "take." Hydraulic paths may become tortuous as a result of the wall roughness or the irregular filling of discontinuities. This may lead to significant head loss, which in turn can result in grout particles settling and clogging the path.

According to Weaver, investigators found that openings as small as 0.007 mm wide per foot can result in permeability on the order of 10^{-4} cm/s (Vaughan 1963). Cracks of this width are not readily groutable because they will become clogged by the cement, which has a maximum particle diameter of 0.2 mm. For this reason, European practice calls for sufficient pressure to open fractures to a width that will allow the passage of grout particles (Lomardi, 1985). This is usually three or four times the U.S. rule of thumb of 20 kPa/m (1 psi/ft) of depth below the ground surface.

Dry, porous rock tends to absorb water out of the grout. The ensuing increase in viscosity and cohesion prevent grout penetration. Remedies to offset the effects of porosity include grouting with a higher water:cement ratio, preceding the injection of cement grout by the injection of silica gel, and prewetting the porous rock (Cambefort 1977; Weyerman 1958). For prewetting to be successful, the rock must become saturated. This may require long durations of holding water pressure on a grout hole.

In the conduct of water tests for grouting operations, the calculated permeability lead to an erroneous value. This is a likely occurrence when the testing is performed over a time period too brief to allow saturation. Weaver also offers a precautionary note regarding the assumption that impermeable rock mass cannot conduct significant quantities of water. He offers the example of shale, which is generally considered as an aquiclude, but may take a considerable quantity of grout where continuous openings exist.

CHAPTER 4

EQUIPMENT

The various types of equipment used for underground grouting are basically the same as that used aboveground. In underground applications, however, the equipment size, layout, and mobility are determined to some degree by the size and shape of the underground structure being grouted. Figs. 4-1 and 4-2 show typical space restriction problems. It can be seen in these examples that the grouting equipment completely blocks access of other equipment and materials around the operation. This is particularly true in smaller openings, of 6 m (20 ft) or less. Additionally, access of grouting equipment and materials into the area being grouted may be the governing factor dictating the equipment sizes and layout to be used. For example, the space available within an underground hydroelectric powerhouse may be more than ample for large grouting equipment, but access into the area may be via a shaft or tunnel, which could prove to be the limiting factor. Even when the size of the access ways are adequate, scheduling their use for deliveries of grouting equipment and material may be controlled by schedule priorities of the larger overall construction project or mine production schedules. Therefore, careful preconstruction planning and equipment selection are critical for a quality, on-schedule, and within-budget grouting program.

This chapter examines the types and sizes of equipment required to drill grout holes, proportion and mix grout materials, and deliver and inject grout into the holes in an underground work environment. The most common power sources for underground grouting equipment are air, electric, and electric-hydraulic. Diesel-powered equipment is also common and, when used, requires the installation of a scrubber on the equipment's exhaust system. The use of diesel equipment may also increase the underground ventilation requirements beyond what is already in place, thereby requiring additional primary or ancillary ventilation capacity to be installed. Gasoline-powered equipment is not allowed by law to be used underground.

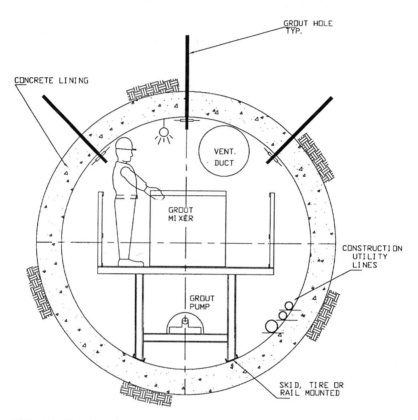

FIG. 4-1. Section of Typical 4 m (12 ft) Finish Diameter Tunnel Illustrating Working Space Restrictions at Mixing and Pumping Operations Area

4.1 DRILLS

The first step in selecting the drilling equipment is to establish the overall scope of the drilling requirements for the project grouting program. These requirements include the total length of grout holes for each grouting method to be used, the diameter and number of holes per method, as well as the working space and location of the various drilling areas.

Depending on the specific grouting methods used, several ranges of hole depth are common:

- 0.3–0.6 m (1.0–2 ft)—contact grouting
- 2–9 m (6–30 ft)—consolidation grouting
- 9–30 m plus (30–100 ft)—curtain grouting
- 10–40 m (30–130 ft)—probing/grouting ahead
- 3–100 m plus (10–330 ft)—shaft pre-grouting

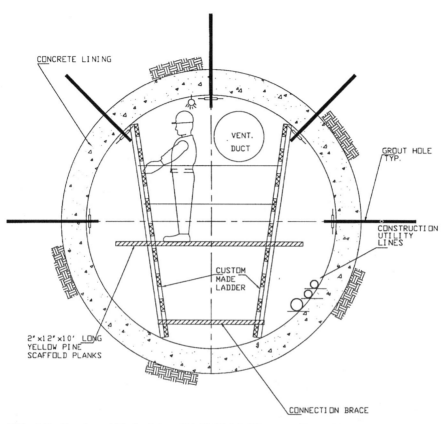

FIG. 4-2. Section of Typical 4 m (12 ft) Finish Diameter Tunnel Illustrating Working Space Restrictions at Mobile Work Deck Used to Access Grout Holes above Spring Line

Two common drilling methods used underground are percussion and rotary. The use of down-hole drills for grouting is less common at present, but their use is increasing in frequency for surface grouting and is expected to increase in underground applications (Continental Drilling, 1993).

Percussion drilling, for underground grouting applications, can be used to effectively drill hole depths up to approximately 18 m (60 ft). Rotary drills are better suited for holes in excess of 18 m (60 ft). These criteria are based on the fact that proper hole alignment becomes difficult beyond 18 m (60 ft) when using percussion drilling. Also there has long been a controversy in the grouting industry as to the effect that the drilling method, that is, percussion versus rotary, has on the groutability of the rock through the drill-hole wall surface. The condition of the wall surface of the hole is a function of the drilling method used. One scenario is that rock chips, produced by the impact or chipping action of percussion drilling could preseal fractures in the rock, thus preventing the injection of slurry grout. This problem can be alleviated by requiring thorough flushing of the hole with water to remove all the cuttings caused by the percussion drilling. On the other hand, because of the grinding action of rotary drilling, presealing of the grout hole is not seen as a problem when rotary drilling is used.

4.1.1 Percussion Drilling

In percussion drilling the drill steel and bit are rotated as a single rigid unit by an air- or hydraulic-powered drill motor. Simultaneous with the rotation, a hammering action is transmitted through the drill steel into the bit. The bit remains in close contact with the rock at the bottom of the hole, producing a chipping action that breaks the rock into small fragments. These fragments, called "cuttings," are removed from the drill hole with compressed air and water.

For drilling the short contact grout holes and consolidation grout holes up to approximately 3–3.6 m (10–12 ft) deep, jack leg, stoper, and sinking drills are generally used. In the industry jack leg drills are also called "air leg" drills, and sinking drills are also called "jack hammers." When the grout hole depth exceeds 3–3.6 m (10–12 ft), other types of percussion drilling equipment are used. In tunnels and chambers exceeding approximately 4 m (12 ft) in height and width, rubber-tired and track-mounted percussion drills are used for drilling holes up to approximately 18 m (60 ft) deep. Once in these larger openings, the mobile drilling equipment can also be used, for speed and efficiency, to drill the shorter length contact and consolidation holes. Percussion drilling is not, however, recommended for hole depths exceeding 18 m (60 ft) because proper hole alignment can not be maintained. Therefore, when holes deeper than 3–3.6 m (10–12 ft) are required in tunnels smaller than approximately 3.6 m (12 ft) in height and width, or in larger openings when hole depths exceed 18 m (60 ft), rotary drills are used.

A jack leg drill (Fig. 4-3) is a pneumatically powered drill that can be operated by one worker. It is the most common type of percussion drill used for drilling shallow, small-diameter, grout holes underground. It is also the type of drill used in the drill-and-blast excavation method to drill blasting holes for small tunnels up to about 3.6 m (12 ft) in diameter. The drill steel is supplied in lengths ranging from 0.6 to 3.6 m (2 to 12 ft) in increments of 0.6 m (2 ft). The drill bits are usually the chisel type with tungsten-carbide inserts, and typically range from 25 to 50 mm (1 to 2 in.) in diameter. Water and air are used to remove drill cuttings from the hole. The drill can be operated using air only for cutting removal, therefore the use of water during jack leg drilling should be made a requirement in the specifications to maximize hole cleaning as well as a means of dust control.

EQUIPMENT 87

FIG. 4-3. Jack Leg Drill

A jack leg drill is best suited for drilling contact and consolidation grout holes located between 200 and 160 degrees in circular and horseshoe-shaped tunnels up to 3.6 m (12 ft) in diameter. Stoper drills, less common on civil projects than on mining projects, are used to drill overhead and can be used to drill grout holes located between 330 to 30 degrees. Sinking drills are used to drill downward and can be used to drill grout holes in the invert of a flat-bottomed structure or in the lower quarter-arches, located between 160 and 200 degrees, of circular tunnels. Fig. 4-4 shows the effective drilling range for, jack leg, stoper, and sinking drills.

Since these three types of drills are limited to hole depths of 2.5–3.6 m (8–12 ft), relatively small rotary drilling equipment is used if deeper holes are required in smaller tunnels, which do not allow the use of the larger mobile percussion drilling equipment. A more detailed discussion of rotary drills is presented in the next section of this guide.

In larger tunnels and chambers, rubber-tired and track-mounted percussion drilling equipment is often used to drill all of the required grout holes to depths of approximately 18 m (60 ft). Most mobile drilling equipment, which is designed specifically for underground use, has the capability of drilling holes radially through a full circle in a vertical plane. However, the size of most underground mobile percussion drilling equipment limits its use to tunnels and chambers larger than 3.6–4.2 m (12–14 ft) in height and width. Often, for economic reasons, the same piece of underground mobile drilling equipment at the project site is used to drill holes for blasting, rock

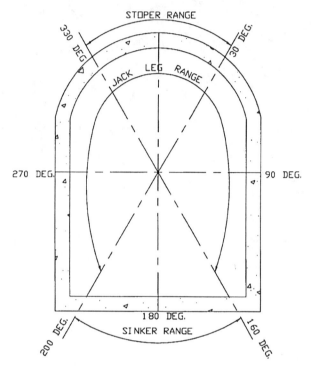

FIG. 4-4. Effective Ranges of Jack Leg, Stoper, and Sinking Drills Used for Drilling Contact and Consolidation Grout Holes

bolting, and grouting. Since excavation and rock bolting precede grouting, the drilling equipment used for grouting may be predetermined on a project by these other drilling activities.

When using percussion drilling, unacceptable levels of hole deviation and alignment problems begin to occur at depths greater than 18 m (60 ft). This is due to the drill bit "wandering." The problem of hole wandering stems from the fact that the drill bit has a considerably larger diameter than the drill rods in order to allow the removal of the drill cuttings. Because of this, the bit tends to deviate from the theoretical centerline of the hole when it hits rock material of different hardness or changes in the geologic structure, or is affected by the downward pull of gravity when holes are drilled at angles other than vertical. This problem can be controlled somewhat by installing centralizers along the length of the drill stem.

The problems of hole deviation and alignment are especially troublesome in underground work because, unlike grouting performed from the surface, all but a few grout holes are drilled at some angle other than vertically downward. The geometry and work environment of underground structures can also add to hole deviation and alignment problems, for example

- Most tunnel grouting is performed in radial rings, therefore, the hole-to-hole spacing at the hole collar, within a circular tunnel, can be very close, on the order of 0.6–1.2 m (2–4 ft). However, because of the radial pattern of the drilling, the hole-to-hole spacing increases with the depth of the hole (Fig. 4-5). Because

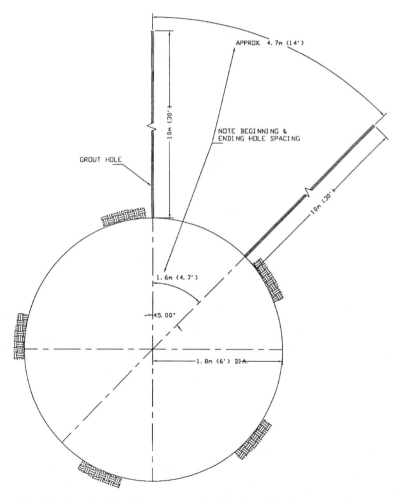

FIG. 4-5. Section Showing Increase in Hole-to-Hole Spacing as Function of Hole Depth

each grout hole has a finite maximum radius of penetration of the grout around the hole, the hole-to-hole spacing is one of the most important factors of a quality grouting program. Therefore hole "wandering," which can significantly increase this spacing with depth, must be kept to a minimum.
- Because most underground grout holes are drilled radially at various angles, it is quite important to set up and align the drilling equipment properly before drilling starts. This critical equipment setup and alignment is further complicated by less than ideal working conditions, such as space restrictions, drilling equipment configuration, lighting limitations, and trying to start a drill hole on the often curved concrete surfaces of lined underground structures.
- There is a limited degree of accuracy of the layout instruments, such as prefabricated triangular templates and carpentry levels, used to set the angles of the drilling equipment.

In view of these difficulties it is expected that hole alignments can deviate within plus or minus 5 degrees from the intended angle at the start of drilling. These compounding factors all affect the hole-to-hole spacing; therefore, they must be considered and compensated for by the designer during the grout program design and specification preparation phase of the project.

4.1.2 Rotary Drilling

Relatively small air-operated rotary drilling equipment is used to drill grout holes in excess of 3–3.6 m (10–12 ft) in tunnels and chambers less than 3.6 m (12 ft) in diameter. The same type of drill is also used in larger tunnels when the grout hole depths exceed approximately 18 m (60 ft) because hole alignment control with rotary drills is superior to that with percussion drilling equipment. Some commonly used rotary drills are the Chicago Pneumatic models CP-55 and CP-65, now manufactured by Longyear and identified as Longyear-55 and Longyear-65. The drill rods and bit are rotated by an air motor that can be operated at one of three rotational speeds by using a gearbox. The drill advances by a screw feed and is mounted on a post or column that allows the drill to be rotated 360 degrees in a vertical plane (Figs. 4-6 and 4-7).

Diamond coring bits are generally used to core the rock of grout holes underground when using rotary equipment. Coring, however, requires the withdrawal of all the drill rods at the completion of each coring run, which are usually 0.3–1.5 m (1–5 ft). The use of a wire-line core removal system is not practical, since most holes are drilled at an orientation other than vertically or near vertically downward. Additionally, since smaller tunnels may only allow the use of 0.3–0.6 m (1–2 ft) long drill rods, their removal

EQUIPMENT

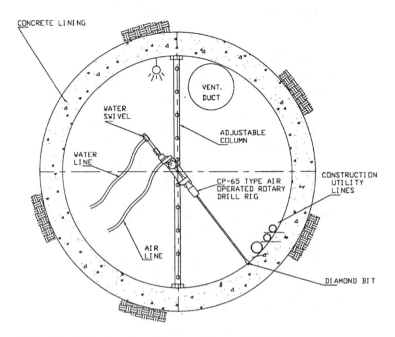

FIG. 4-6. Typical Column-Mounted Rotary Drill Rig

and reinsertion for every core run can become time consuming. This can add considerable cost and time to the overall grouting schedule. An example when this requirement would become a timely process would be when drilling a ring of 20 or more curtain grout holes to depths of 12 m (40 ft) in a 2.4 m by 2.4 m (8 ft by 8 ft) tunnel. In such a case a second drill may be needed to to complete a single ring.

For these reasons some contractors prefer to use diamond plug bits rather than coring bits to drill deeper holes. Plug bit drilling has a slower advance rate than coring because the full rock volume of the hole is ground up in the drilling process, as opposed to only the circumference being cut by a coring bit. A plug bit, however, does not produce a core to remove, so it is not necessary to pull the drill rods from the hole until the full depth of the hole is reached. Because of the fine granular nature of the drill cuttings produced by rotary diamond drilling, only water is used to remove drill cuttings from the hole. Air is not required.

This type of rotary drilling equipment is small enough to be transported and set up by two workers. The drill itself can easily fit into a wheelbarrow for transport or can be carried by workers from one grouting location to another along the tunnel alignment.

FIG. 4-7. CP-65 Column-Mounted Rotary Drill Rig

4.1.3 Down-Hole Drill (Down-Hole Hammer) Drilling

In the down-hole drill, also called the down-hole hammer, drilling method, the drill rods, hammer, and bit, located at the bottom of the hole, are rotated by a drill motor located on the main body of the drill. In this arrangement, unlike percussion drilling, the hammering action takes place at the bottom of the hole where the chipping work is done, thus the name down-hole drill. The hammer, which produces the up-and-down piston action, is powered by high-pressure air. The theory is that because the hammering force is delivered at the bottom of the hole, where the work is done, there is no energy lost through the drill rods, thus maximum hammering force is applied to the rock. In standard percussion drilling, much of the hammer energy needed to chip the rock is lost as it is transmitted through the drill rods. The energy loss increases as the depth of the hole increases.

Down-hole drills are gaining wider use in grouting applications performed from the surface, such as grout curtain drilling for dams, where holes are relatively deep (Continental Drilling, 1993). Presently, however, their use is not widespread underground. This may be due in part to the fact that most grout holes drilled underground are relatively short and down-hole drilling becomes more economical as drilling depth increases. Also, down-hole drills are not presently in common use for other underground drilling applications; therefore, the technology is not well known to the underground industry.

4.2 MIXERS

Two types of mixers predominate in underground grouting applications, paddle mixers and colloidal mixers. The paddle mixer is the older of the two and is still used more frequently underground on smaller projects than the colloidal mixer. This is especially true for the contact grouting of tunnel and shaft linings and for filling the large voids frequently found in naturally occurring underground caverns and cave systems. In these applications the complete wetting and dispersion of the cement particles in the grout mix, a shortcoming of the paddle mixer, is not as critical as when performing consolidation and curtain grouting. Also, in contact, cavern, and cave applications, short interruptions of grout injection due to a slower mixing rate of paddle mixers are more tolerable.

A paddle mixer consists of either vertically or horizontally mounted paddles that rotate slowly within a cylindrical tank. During the mixing process, the grout materials are stirred and thrown against baffles attached to the side of the tank. The baffles prevent the formation of a vortex within the swirling mixture, which can cause centrifugal separation. The heavy materials are thus better broken apart and more evenly distributed throughout the mixture. Once completely mixed, the grout is discharged from the mixer into an agitator tank when a single-mixer arrangement is used or directly into the grout pump when using a twin-mixer arrangement (Figs. 4-8 and 4-9). In either arrangement, grout should be discharged by direct suction, not gravity, into the grout pump.

Gravity discharge into the pump should be avoided because it can potentially allow air to enter the pump. The presence of air in the grouting system can cause erroneous pressure readings. Gravity discharge still, however, remains a popular method. Therefore, the use of a direct suction system should be made a requirement in the specifications.

Paddle mixers range in capacities from approximately 283 L (10 ft^3) to 1,417 L (50 ft^3) and larger. The size and arrangement of mixer(s) chosen depends on the amount of grouting expected, the type of grouting to be performed, and the size of the working space available for the equipment.

The high-speed "colloidal" mixer was introduced in the United States in 1955 by the Colcrete Company of Strood, England (Weaver 1991). It works by recirculating the grout through a centrifugal pump. The pump, which

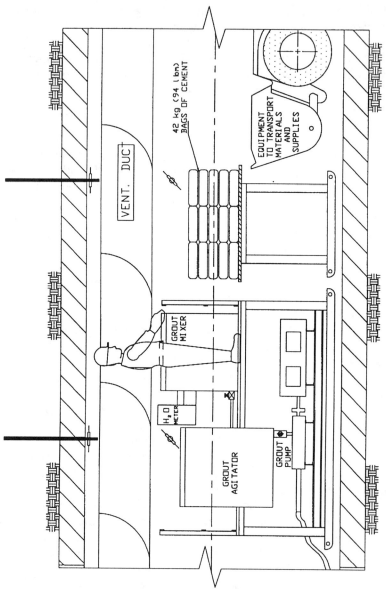

FIG. 4-8. Longitudinal Section of Typical 4 m (12 ft) Finish Diameter Tunnel Showing Single-Mixer and Agitator Arrangement

EQUIPMENT

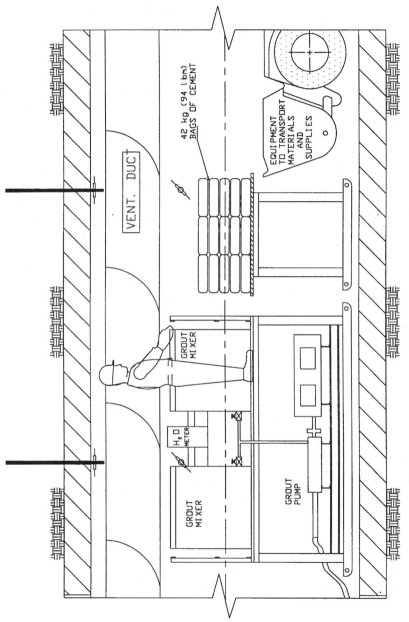

FIG. 4-9. Longitudinal Section of Typical 4 m (12 ft) Finish Diameter Tunnel Showing Two-Mixer Arrangement

operates at high speeds, of 1,500–2,000 revolutions per minute, imparts a shearing force to the grout as it passes through the narrow steel pump casing. The pump discharges the grout tangentially into a vertical cylindrical tank causing a vortex to form (Figs. 4-10 and 4-11). The centrifugal force created throws the heavier, unmixed grout fraction of grout against the tank walls, where it then runs down the wall and reenters the pump for further mixing.

The speed of mixing and the high output capacity are important attributes of colloidal mixers. The time required to produce a thoroughly mixed batch of grout is reported to be only 15 sec (Water Resources Commission of New South Wales 1981). Colloidal mixers range in capacities from approximately 283 L (10 ft^3) to 14,170 L (500 ft^3) and larger. As with the paddle mixers, the size and arrangement chosen depends on the amount of grouting expected, the type of grouting to be performed, and the size of the working space available for the equipment. Fig. 4-12 shows a complete rail-mounted grout plant consisting of colloidal mixer, agitator, and grout pump. The mixing action of the colloidal mixer produces grout with superior properties compared to grout mixed in paddle mixers. The current trend is to specify colloidal mixers rather than paddle mixers in contract documents for all grouting, but particularly for consolidation and curtain grouting.

Most grout mixing is performed relatively close to the injection location; other equipment arrangements are, however, sometimes employed. For example, a large colloidal mixing plant might be set up at the top of a shaft or just outside the tunnel portal. From these aboveground plants the thoroughly

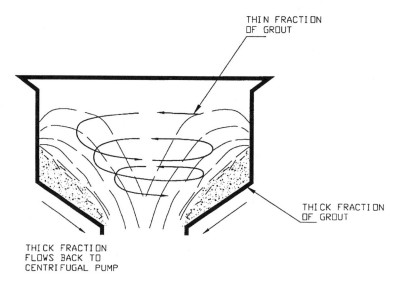

FIG. 4-10. *Vertical Cylindrical Tank of Colloidal Mixer Showing Formation of Vortex*

FIG. 4-11. Colloidal Mixer

FIG. 4-12. Rial-Mounted Colloidal Mixer with Agitator and Grout Pump

mixed grout is pumped underground through a system of pipes. The grout is discharged into an agitator tank that in turn feeds the grout pump near the point of injection (Fig. 4-13).

A standard concrete batch plant can also be used to proportion and mix the grout at an aboveground location. The thoroughly mixed grout can then be discharged from the plant directly into a specially designed piece of equipment called a "Moran car." A Moran car is a horizontal concrete or grout agitator tank that rides on a railroad track. It is used to transport concrete and grout underground (Fig. 4-14). Moran cars are sized to carry between 4.5 and 9 m^3 (6 and 12 yd^3) of grout. From the Moran car the grout is transferred into a standard grout agitator and pump arrangement located close to where the grout is to be injected. Rubber-tired versions of the Moran car are also used to transport grout underground. Sometimes, depending on the location of the batch plant, it is necessary to transfer the grout from the batch plant into a concrete transit mix truck and than into a Moran car.

Grout from concrete batch plants can also be delivered down drop-pipes, which are small diameter shafts or drill holes installed to deliver grout and concrete to underground locations. It is also common for grout to be discharged from a transit mix truck into pumps at the top of the shaft or tunnel portal and pumped via pipelines to the point of injection.

4.3 AGITATORS

An agitator is a storage tank where the thoroughly mixed grout from the mixer is stirred by a slowly revolving paddle to keep the particles of unstable grout in suspension while awaiting injection. Since the stirring action alone cannot keep the denser particles of grout from settling out of the mixture, baffles are attached to the sides of the tank to help produce turbulence, thereby reducing settlement during agitation. Some agitator manufacturers create this turbulence by mounting the paddle at a diagonal to the vertical axis of the tank (Fig. 4-15).

Agitators range in capacities from approximately 283 L (10 ft^3) to 14,170 L (500 ft^3) and larger. It is important to size the agitator capacity to about 25% greater than the mixer capacity. This allows the entire contents of the mixer to be discharged into the agitator before the agitator runs dry, thus avoiding interruptions of the grout injection and the possible ingestion of air into the pump. The discharge port from the mixer to the agitator, as well as the discharge opening of the grout return line, which returns unused grout back into the agitator, should be equipped with a mesh screen. The screen filters out lumps of unmixed grout, pieces of cement bags, scale from grout lines, and other foreign material. If not removed, these contaminants can clog the grout hole, thereby causing premature refusal and interruptions to the grouting operation.

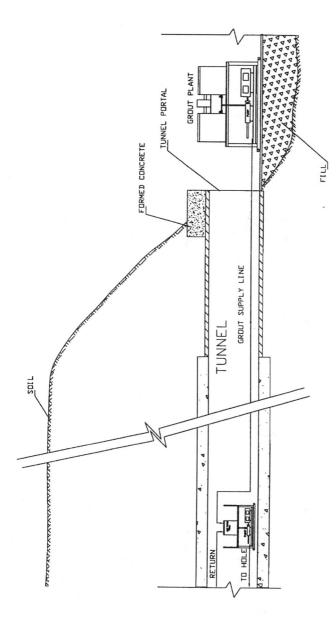

FIG. 4-13. Large Colloidal Mixing Plant Set Up outside Tunnel Portal

FIG. 4-14. Moran Car

4.4 WATER METERS

Water meters for measuring the amount of water added to the grout mix should always be required by the specifications. Other methods, such as calibrated buckets and predetermined water level markings on the sides of the mixing tank should not be allowed. Water meters calibrated in United States gallons, Imperial gallons, liters, cubic feet, and cubic meters are all available.

A common method of proportioning grout for underground applications in the United States is by volume, especially on small grouting projects. When using the volumetric method, the water meter should be calibrated in cubic feet. Water meters calibrated in U.S. gallons should be avoided because the quantity calculation can become confusing to the mixer operator. This is especially true when it is necessary to change the water:cement ratios midway through grouting a hole. When grout quantities are expected to be large, high-capacity batch plants are used. These high-capacity plants generally proportion the grout materials, including water, by weight rather than volume.

Water meters are available for water-line sizes from 16 mm (5/8 in.) to 38 mm (1.5 in.). The most common size of water line used underground is 25 mm (1 in.). Use of water lines smaller than 19 mm (3/4 in.) should not be allowed. The use of smaller-size lines cause an increase in the time required to charge the mixer with water. This in turn can cause an interruption of the grout injection.

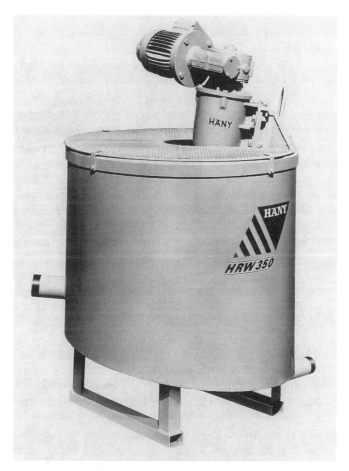

FIG. 4-15. Diagonally Mounted Paddle Agitator

A water meter with a "reset-to-zero" feature is strongly recommended. This eliminates the risk of the mixer operator miscalculating the correct cumulative meter reading required for the proper amount of water to be added in successive batches. Meters are also available that can be preset to automatically shut off when the desired amount of water has been discharged into the mixer. A large, easy-to-read dial face is important, especially with the less-than-ideal lighting conditions found underground.

4.5 PUMPS

Two common pumps used for grouting underground are the progressing helical cavity pump and the piston pump. The progressing helical cavity pump is, however, the one used most often. It is often referred to as a "Moyno pump" in the industry, although Moyno is a brand name. Many other manufacturers supply progressing helical cavity pumps.

Progressing helical cavity pumps produce a continuous, uniform flow of grout into a hole at relatively constant pressure. Most progressing helical cavity pumps used underground are direct drive, that is, they have no gearbox, and are air powered. The pump speed and grout output are controlled by using a valve to regulate the amount of air delivered to the drive motor. The quantity and pressure of grout entering the grout delivery line can be further controlled with a grout bypass line. By using a valve to control flow, the bypass line returns grout from the pump discharge back into the mixer or agitator before it enters the grout delivery line. When a diesel engine is used to power a progressing helical cavity pump a standard shift automobile-type transmission is used to control pump speed. This in turn controls output and pressure. The bypass line return arrangement is also used to further control the amount of grout and pressure entering the delivery line.

Progressing helical cavity pumps are used primarily to pump grout mixtures of water, cement, and bentonite; however, they are also capable of pumping sanded grout mixes. The abrasiveness of the sand, however, increases the wear on the pump. Grouting contractors using progressing helical cavity pumps are, therefore, often reluctant to switch to a sanded mix even when field conditions clearly dictate its use.

The unexpected need to pump large quantities of sanded mixes or the prolonged use of sanded mixes on low volume grout projects may generate a claim for additional time and money by the contractor. Therefore, if the possibility exists that a sanded mix may be required, this fact should be made clear in the bidding documents. In an effort to avoid a claim situation, it is important to clearly outline in the contract documents the design assumptions relating to the grout mixes and the relative quantities expected. If, for example, the grouting of large voids requiring the use of sand as an economic filler is anticipated, this should be clearly stated in the contract documents. This will allow the contractor to better plan and schedule the allocation of equipment for the project.

If the use of sand is anticipated, the contractor may elect to have a piston pump on site. A piston pump is better suited to pump sanded mixes. It is also a good practice for the contract documents and the field inspection plan to be written to allow maximum flexibility in modifying the grout program and equipment to best suit the actual field conditions encountered.

Piston pumps are used predominantly to fill large voids and caverns and for backfill grouting behind precast and steel liners. They are also used, to a lesser degree, for contact grouting. One disadvantage of piston pumps is that they deliver a pulsating pressure that makes pressure control difficult when constant or low maximum pressures are required. Another disadvantage of the piston pump is it occasionally becomes plugged during injection, which can result in the loss of the hole being grouted (Weaver 1991).

4.6 PRESSURE GAUGES

Pressure gauges are used to monitor the injection pressure of the grout being delivered to the grout hole. The pressure gauge helps to insure that the maximum allowable pressure is being applied to achieve the desired results of the grout program design. It also helps to prevent the application of an injection pressure that is too high and could overstress the foundation materials and the structures being grouted.

In the United States, pressure gauges are calibrated in pounds per square inch (psi), whereas the Système Internationale (SI) measures pressure in bars. The face of the gauge should be a minimum of 75 mm (3 in.) in diameter with 100–150 mm (4–6 in.) preferred. This size requirement is important, given the less-than-ideal lighting conditions found underground. The minimum allowable face diameter for the gauge should be specified.

The range of pressure reading of the gauge should be appropriate for the grout pressures being used. For example, a gauge with a pressure range from 0 to 3.5 bars (0 to 50 psi) would be appropriate for contact grouting steel and concrete tunnel linings. A gauge with a pressure range of 0–8 bars (0–120 psi) would be appropriate for consolidation grouting holes up to 9 m (30 ft) deep.

Gauges are easily damaged in the underground construction environment; therefore, they should be checked for calibrations at least once per week or more often when necessary, based on any erratic pressure reading or physical damage to the gauge. A master gauge used to check calibrations of the production gauges should be maintained on site in a protective storage box within the field office. This gauge should never be used for production grouting.

4.7 GAUGE SAVERS

The grout mixture should never be allowed to come into direct contact with the pressure gauge. Therefore, a protective medium must be used to separate the grout from the gauge. A gauge saver, also called a "diaphragm seal," can be used for this purpose (Fig. 4-16). Within the body of a gauge saver the upper portion is isolated from the grout by a diaphragm. The area between the top of the diaphragm and the pressure gauge is completely filled with a suitable fluid, usually oil. Displacement of the fluid through movement of the diaphragm transmits pressure changes to the pressure gauge. Other methods to achieve this grout/gauge separation are available (Fig. 4-17); however, the diaphragm gauge saver is the most common.

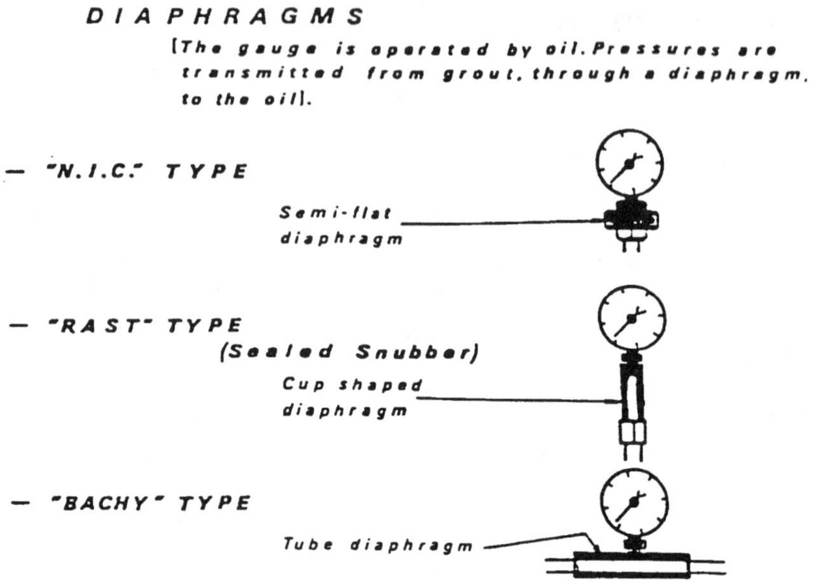

DIAPHRAGMS

(The gauge is operated by oil. Pressures are transmitted from grout, through a diaphragm, to the oil).

— "N.I.C." TYPE

    Semi-flat diaphragm

— "RAST" TYPE (Sealed Snubber)

    Cup shaped diaphragm

— "BACHY" TYPE

    Tube diaphragm

FIG. 4-16. Diaphragm Gauge Savers (Water Resources Commission of New South Wales 1981)

U-TUBE

This is a simple, reliable arrangement, provided that care is taken to
— keep the "U" hanging downwards.
— make sure that the oil is not accidentally drained out when moving gauges and fittings around the job.

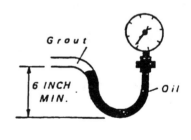

FIG. 4-17. U-Tube Gauge Saver (Water Resources Commission of New South Wales 1981)

4.8 PACKERS

Packers are used to seal off or isolate a portion of a grout hole that allows grout to be injected under pressure into a specific section of the hole. Packers are installed either at the top of the hole, also called the hole collar, or at other locations along the length of the hole. In underground grouting applications the packer is set at the top of the hole for contact grouting. It is also most often installed at the top of the hole for consolidation grouting operations when the zone of rock to be grouted is directly beyond the excavation limits, say within 3–4 m (10–12 ft). Figs. 4-18 and 4-19 show mechanical packers installed in steel grout pipes that were installed into a shotcrete tunnel liner. The grouting in the photographs was performed in a mixed-face tunnel to control groundwater entering the excavation through sand located in the upper third of the tunnel cross section. When the zone of rock to be consolidation grouted is at a greater depth, the packer is set just above the zone to be grouted. When curtain grouting, however, the packer is usually set at several points along the hole in addition to the top. The stage-up or stage-down grouting methods are normally used. The various grouting procedures used in curtain grouting are discussed more fully in Chapter 8.

Holes are grouted using a single-packer or double-packer arrangements. In a single-packer arrangement, the one most used in underground applications, the grout is discharged into the grout hole just below the location of the packers. In the double-packer arrangement, the grout is discharged into an isolated section of the grout hole located between an upper and lower packer.

The single-packer arrangement is used in contact and most consolidation grouting applications. The double-packer arrangement is used to grout isolated sections such as a water-bearing zone in both consolidation and

FIG. 4-18. Mechanical Packers Installed at Hole Collar through Shotcrete Liner

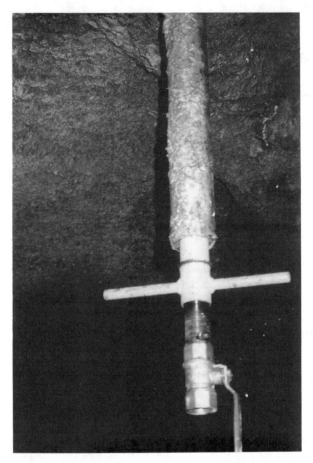

FIG. 4-19. Close-Up of Mechanical Packer Shown in Fig. 4-18

curtain grouting operations. Two common types of packers are the mechanically activated packer (Fig. 4-20) and the air-inflatable packer (Fig. 4-21). In addition to compressed air, high-pressure bottled gasses such as nitrogen and hydraulic fluids are used to inflate packers when high-pressure grout injection is performed. Mechanical packers are the most common type used underground.

The typical hole diameters used for underground grouting range between 32 and 63 mm (1.25 and 2.5 in.). Packers are commercially available for these ranges and are normally available in diameter increments of 6 mm (1/4 in.).

The length of the packer assembly used depends on such factors as the frequency and orientation of the joints and fractures in the rock, the maxi-

EQUIPMENT 107

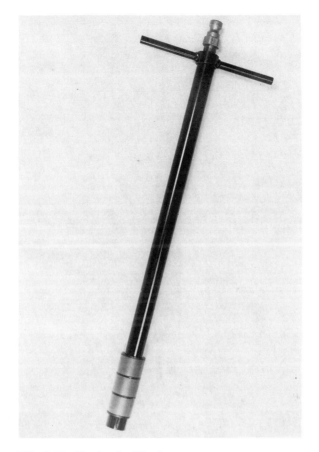

FIG. 4-20. Mechanical Packer

mum injection pressure used, and the texture of the walls of the drill hole. The hole's wall texture is governed by the drilling method used, that is, percussion or rotary, and the type of rock encountered. Common packer lengths used are between 0.3 m (1 ft) and 1.0 m (3 ft). Fig. 4-22 shows a sketch of a 0.3 m (1 ft) mechanical packer installed in a concrete tunnel liner to allow for contact grouting between the liner and the rock.

4.9 NIPPLES

Nipples are short lengths of pipe used to connect the grout delivery lines to the grout hole. To facilitate attachment, a pipe union is embedded in the precast concrete liner or welded directly to the steel liner. This allows the pipe nipple to be screwed in and attached at the collar of the grout hole.

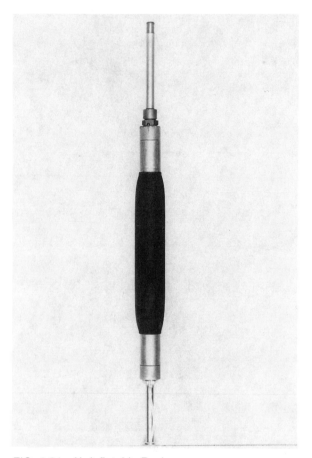

FIG. 4-21. Air-Inflatable Packer

Nipples can also be installed directly into holes drilled in cast-in-place liners or in rock (Fig. 4-23). Nipples can be used in lieu of a packer when grouting is performed from the top of the hole. The use of nipples to grout behind steel liners is discussed in more detail in Chapter 2, subsection 2.3.2, embedment grouting.

Most contractors prefer not to embed pipe unions in cast-in-place concrete because layout and installation of the unions onto the concrete formwork can be a time- and labor-intensive process. There is also a high risk that the unions will become dislodged and move during the concrete placement. Unions are, however, commonly installed in precast liner segments because their layout and installation can be better controlled in the precast plant environment.

EQUIPMENT

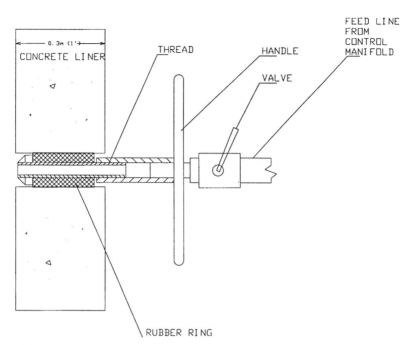

FIG. 4-22. Mechanical Packer Arrangement for Contact Grouting of Void between Rock and Concrete Liner

4.10 DELIVERY AND DISTRIBUTION SYSTEM

The delivery and distribution system consists of all the grout lines, valves, gauges, and fittings between the discharge end of the pump and the hole being grouted.

The delivery, or supply, line brings the grout from the pump to the header or control manifold. The return line recycles the unused grout back to the agitator tank or mixer. The return line also serves as a pressure relief line to help control the pressure at the header. The use of delivery and return lines less than 25 mm (1 in.) in diameter should not be allowed because of the danger of plugging the lines.

The header is the assembly where the operator controls the grout injection pressure and flow rate (Fig. 4-24). The header assembly should be sized so that it can be carried as a unit by the workers from one grout injection point to the next.

Diaphragm valves, also called "Saunders" valves, are the most suitable valves (Fig. 4-25) for grouting. They can be used for grouting pressures below 14 bars (200 psi). The valve body should be made of steel and can be supplied with a rubber liner. Lubricated plug cocks can be used for pressures

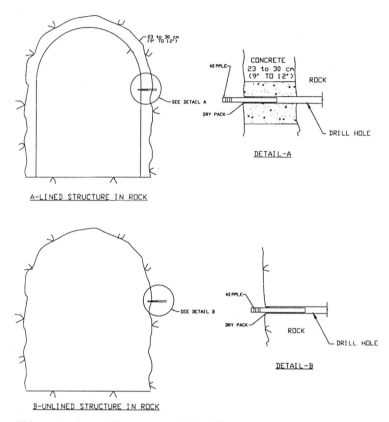

FIG. 4-23. Grout Nipples Installed in (A) Lined and (B) Unlined Underground Structures

above 14 bars (200 psi). Unlubricated plug cocks and ball valves can be used for the bleed-off and packer valves (Fig. 4-26).

4.11 DATA ACQUISITION AND RECORDING EQUIPMENT

The data acquisition and recording equipment consists of measuring units and the recording unit. The measuring units measure flow rate and pressure. In the recording unit the electronic signals of the measuring unit are processed. The flow rates and pressures are displayed digitally and, if desired, a permanent paper copy record is produced. The equipment can be used to measure and record data from both grouting operations with cement-based grouts and permeability testing.

The flow rate is measured with a inductive flow meter that measures the flow electronically and has an unobstructed passageway with no moving

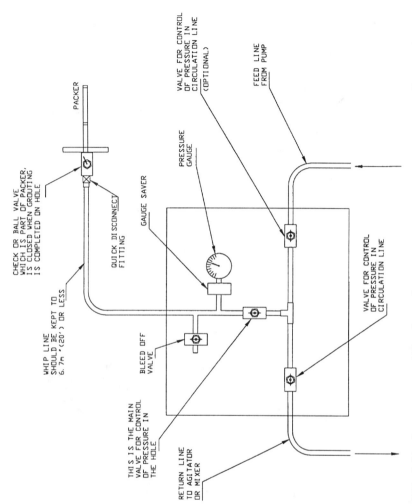

FIG. 4-24. Arrangement of Header or Control Manifold Used to Control Flow Rate and Pressure at Grout Hole [Header Located within 3 m (10 ft) of Hole Being Grouted]

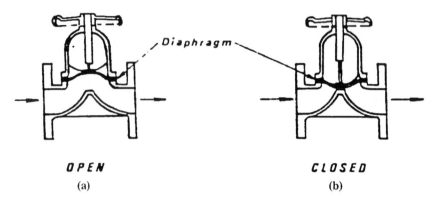

FIG. 4-25. Saunders Valve (Water Resources Commission of New South Wales 1981)

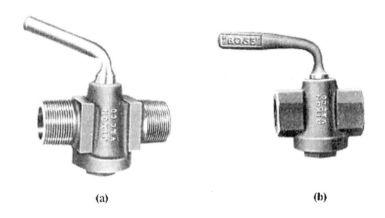

FIG. 4-26. Typical Bleed-Off and Packer Shutoff Valves (Courtesy of Dixon Valve & Coupling Company 1995)

parts. Manufacturers of the meters claim the composition of the grout has no influence on the measuring results provided that some specified minimum conductivity of the grout is maintained.

The pressure is measured using a transducer that is separated from the grout by a diaphragm in order to protect the sensitive instruments. The pressure should be measured as close as possible to the point of injection. This gives a more accurate reading of the actual pressure being applied to the grout hole and precludes inaccuracies in pressure readings caused by friction losses developed between the grout pump and the point of injection.

A shielded cable is used to transmit the electronic signal from the pressure transducer to the recording equipment. Shielded cable length of up to 300 m (1,000 ft) can be used. There are also pressure transducers available to measure pressure in the hole at the packer location. These instruments are called down-the-hole pressure transducers. Their use for grouting is not recommended because a high level of experience and care is needed by the field crew to properly install and maintain the equipment. This level of experience is usually not found among grout crew personnel. The down-the-hole pressure transducer is, however, more widely used for permeability testing when experienced, technically trained personnel are performing the equipment installation, maintenance, and testing.

Recording units are available that display flow rates in L/min or gpm with the accumulated totals displayed in liters or cubic feet. The pressure can be displayed in bars or psi. Some recording units on the market can be preset to a maximum pressure and or maximum grout quantity. When reaching the preset pressure and or quantity, the grout pump is automatically switched off by a potential free contact. Fig. 4-27 shows a recording system. The system shown is a HANY HIR 001. As a frame of reference Table 4-1 shows some technical data for the HIR 001.

4.12 AUTOMATED BATCHING SYSTEMS

On projects where use of large quantities of grout is anticipated, the use of a high-volume automated batching system may be warranted. The automated batch plant should be located on the surface as close as possible to the shaft collar or portal opening. Good access to the plant must be provided and maintained for bulk deliveries of grouting materials.

Complete, automated batch plants are available as preassembled units; Fig. 4-28 shows such a preassembled plant. The plant shown is a Hany MCM 6500 Portable Mixing Plant. This type of plant is designed to store, proportion, mix, and pump suspension-type grout mixes. The Hany unit shown and others on the market are capable of batching grouts consisting of water and up to three dry components. The dry components are stored, in bulk, in silos located alongside the plant. After the water has been added into the mixer, the dry components are fed, one by one, from the silos into the mixer by screw conveyors. The amount of each component added is controlled by a sophisticated weighing system located in the mixer. The colloidal mixer of the unit is used for mixing the grout. When mixing is complete, the grout is transferred into a holding tank that feeds the grout pump. The operation of the entire plant is automatically controlled by the rate of grout consumption. For example, the level of grout in the holding tank is the determining factor whether a new batch of grout is mixed or not. Likewise, if the mixing process has already started, the level in the

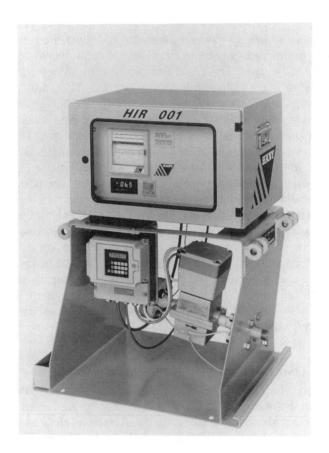

FIG. 4-27. HANY HIR 001 Recording System

holding tank will determine automatically when the mixer is discharged. If the grout is required to remain in the mixer for an extended period of time before being discharged, the automated control system will periodically stop and start the mixing process. This will help avoid heating up of the grout mix while keeping all the components in suspension. These types of batch plants can also be operated manually.

The various functions of the plant are controlled by an automated control unit, Figs. 4-29 and 4-30 show such units. The units shown can be programmed, using the input keyboard, to weigh out up to four components (water and three dry components). A maximum of 10 different grout recipes can be stored by the unit. The display face of the unit shows a diagram of the plant with lights that indicate all the functions. The control unit can be equipped with a printer that produces a printout of the recipe used by

EQUIPMENT

TABLE 4-1. HIR 001 Technical Data

Voltage: Standard	220 V, single phase, 50 hz
Optional	110/440V, 60 hz
Pressure rating: Standard	PN 40 bar (580 psi)
Optional	PN 100 bar (1,450 psi)
Minimum Conductivity of liquid	10 μs
Measuring range:	
Flow: Standard	0–100 L/min (0–26 gpm)
Optional	0–200 L/min (0–53 gpm)
Pressure: Standard	0–40 bar (0–580 psi)
Optional	0–100 bar (0–1,450 psi)
Connections	1 in NPT
Length	750 mm (2 ft 5 1/2 in.)
Width	570 mm (1 ft 10 3/8 in.)
Height	900 mm (2 ft 11 1/2 in.)
Weight	95 kg (209 lb)

FIG. 4-28. Preassembled Grout Plant

FIG. 4-29. Automated Control Unit

identification number, quantity of each component added to the grout mix, and total quantities of each grout recipe batched. Hany's manufacturer's data for the MCM 6500 plant claims mixing capacities of 30–50 m³/h (39–65 yd³/hr), depending on the composition of the mix.

The grout is pumped from the plant via the pipeline to an area close to the point of injection. The grout is discharged as needed from the pipeline into an agitator and a grout pump equipment arrangement near the point of injection. If a pipeline transport system is not used, the grout leaving the plant can also be discharged directly into a Moran car, transit mix trucks, or some other similar piece of equipment for transport to the injection area (see Section 4.2, Mixers, for more detail of these modes of transport).

EQUIPMENT

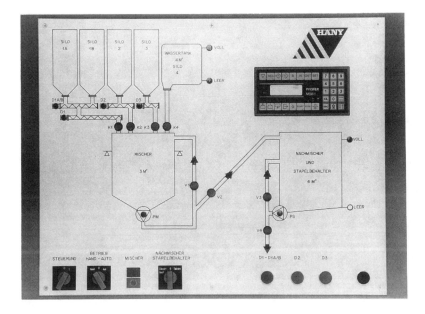

FIG. 4-30. Illuminated Flow Diagram of Automated Control Unit

4.13 GROUTING JUMBOS

A grouting jumbo is a custom-fabricated, mobile structure usually made of steel. It is used as a working platform to allow workers to perform grouting at high elevations within the underground structure being grouted. Most grouting jumbos are fabricated new or are used equipment modified for specific projects. In larger underground openings the jumbos are designed to allow rubber-tired or rail-mounted equipment to travel unimpeded through the center of the jumbo (Fig. 4-31). In smaller tunnels, the jumbo will block access around the grouting operation (Figure 4-32).

Conventional commercially available equipment such as scissor lifts (Figs. 4-33 and 4-34) are also used as grouting jumbos. This type of equipment, however, depending on the size of the underground opening, may not allow for the passage of other materials and equipment beyond the grouting operation.

4.14 EQUIPMENT CONFIGURATION

A typical grouting equipment configuration within a tunnel is shown in Fig. 4-35. The distance from the mixer and pump location to where the grout is injected into the hole should be kept to no more than approximately 15 m (50 ft). A clear, unobstructed line of sight should be maintained at all times

FIG. 4-31. Grouting Jumbo Designed to Allow Other Equipment to Pass through Its Center

FIG. 4-32. Rail-Mounted Grouting Jumbo with Fold-Out Upper Deck

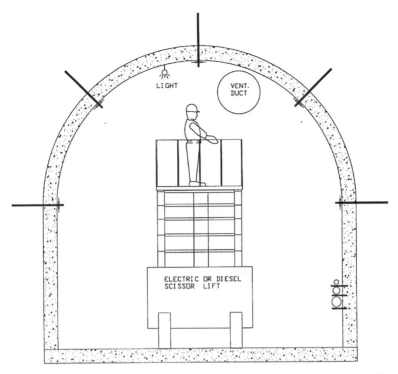

FIG. 4-33. Section Showing Scissor Lift Used as Mobile Work Deck for Access to Grout Holes above Spring Line in Smaller Tunnels

FIG. 4-34. Large Truck-Mounted Scissor Lift Used in Large Diameter Tunnels an Chambers

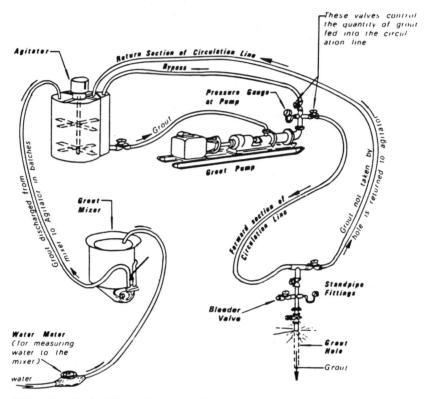

FIG. 4-35. *Typical Grout Equipment Configuration (Water Resources Commission of New South Wales 1981)*

between the pump and header operators. The noise levels around the grouting operation are high; therefore, verbal communication is difficult, especially beyond a few feet. A hard-wired underground communication system with headphones (headset) should be used by the grout crew. The use of a hard-wired telephone system to communicate between the grout operation and the surface should also be used. The telephone system can be used to order materials, for maintenance, and in cases of emergency.

CHAPTER 5

MATERIALS

The basic materials used to produce grouts for underground use are the same as those used in aboveground applications. There is, however, a tendency to keep the mixes basic. This is particularly true in contact and consolidation grouting. Neat grout comprised of portland cement and water are commonly used underground. However, mixes containing bentonite, a natural clay used primarily to reduce settlement of the cement particles in the grout, are recommended. When grout takes are high or known open joints, voids, and cavities must be filled, a sanded grout mix should be used. The addition of sand as a filler to a grout mix is done to reduce the material costs and to help minimize shrinkage of the hardened grout. Ultrafine cement is sometimes specified where a smaller cement particle size will help increase the grout penetration when grouting fine graded soils, tightly jointed rock, and other very small openings. Ultrafine cement grouts are often specified for use in grout curtains. Other fillers and admixtures that modify the characteristics of the grout are also used. Historically, however, their use has not been widespread underground. More recently the trend has been to use the advances made in admixture technology, developed for the concrete and surface grouting industries, to underground applications.

5.1 PORTLAND CEMENT

Portland cement is a hydraulic cement composed primarily of hydraulic calcium silicates. It is the most common cement used in cementitious grout. Hydraulic cement sets and hardens by reacting chemically with water. This process is called "hydration." Concrete and grout made with portland cement will set and harden underwater.

Manufacturers produce several types of portland cement to meet various physical and chemical requirements for their specific uses. ASTM Designation C150, "Standard Specification for Portland Cement," provides for eight types of portland cement as follows:

- Type I: normal, a general purpose cement for all uses where the special properties of the other types are not required
- Type IA: normal, air-entraining
- Type II: moderate sulfate resistant, used where precaution against moderate sulfate attack of the concrete or grout is important
- Type IIA: moderate sulfate resistant, air-entraining
- Type III: high-early strength, provides rapid strength gain
- Type IIIA: high-early strength, air-entraining
- Type IV: low heat of hydration, used where the rate of temperature rise and amount of heat generated from hydration must be minimized. This property is usually not important in grouting, as may be the case in mass concrete.
- Type V: high sulfate resistance, used when the concrete or grout is exposed to severe sulfate action

A Type II, moderately sulfate resistant, portland cement is most often specified for grouting underground. Since there is a high probability that the grout will come into contact with groundwater or soils containing sulfates, the moderate sulfate resistance properties of Type II cement is a desirable attribute.

The type of cement selected is sometimes based on the particle size of the cement as it relates to the sizes of the openings being grouted. For example, Type III portland cement has a much smaller particle size, having a Blaine fineness of 540 m^2/kg, whereas Types I and II cement have a Blaine fineness of 370 m^2/kg. A Type III cement may thus be more appropriate for grouting a tightly fractured rock or a fine graded soil.

On small- to medium-sized grouting projects in the United States, most cement used underground is supplied in 42.6 kg (94 lb) paper bags. Since underground grouting operations are relatively mobile, the use of bagged cement makes it easier to move the cement supply from one grouting location to another. The bags are usually shipped and transported underground on wooden pallets. A typical pallet contains 35 bags of cement. When using the volumetric method of proportioning the grout mixture, a bag of cement is considered to be 0.03 m^3 (1 ft^3). Portland cement can also be supplied in bulk. Bulk cement can be delivered to the project site in 454 kg (1,000 lb) or 907 kg (2,000 lb) plastic-lined, high-strength nylon bags. Bulk cement may also be delivered in much larger quantities by over-the-road tanker trucks or railroad tanker cars. The use of a bulk cement supply is usually associated with a larger aboveground grout plant or an on-site concrete batch plant used for mixing grout. When plants are used to batch grout, the cement and other ingredients of the mix are proportioned by weight.

5.2 ULTRAFINE CEMENT

Ultrafine cement is a portland cement or slag with an average particle size of 4 microns or less. These finely ground materials are also called microfine and superfine cements. All of these terms should be considered synonymous, since at present there is no industry standard.

The use of grout made with ultrafine cement for underground grouting of rock is usually limited to consolidation grouting of tightly jointed rock and grout curtains installed in association with hydraulic structures where superior permeation and penetration of the surrounding rock by the grout are required. Grouts made with ultrafine cement are also used for embedment grouting to fill very small voids, usually caused by shrinkage, between steel shaft and tunnel liners and the backfill concrete or grout. Additionally, ultrafine cement grouts are used in grouting fine-grained soils.

Presently there is no established nomenclature nor reference standard in terms of grain-size distribution for ultrafine cements. The American Concrete Institute (ACI) Committee 552 (Geotechnical Cement Grouting) has unofficially adopted a reference standard for ultrafine cement. In a draft report on state-of-the-art geotechnical cement grouting this committee defines ultrafine cement particles as 100% finer than 15 microns.

There are several brands of ultrafine cements on the market. One such product distributed by the Geochemical Corporation of New Jersey is MC-500 Microfine Cement. It is made with 75% slag and 25% portland cement. It has a Blaine fineness of about 9,000 cm^2/g with 50% of the particle size less than 4 microns. Geochemical Corporation also distributes MC-300 Microfine Portland, which is a portland cement–based product. It has a Blaine fineness of about 12,000 cm^2/g with 50% of the particle size less than 3.5 microns. Another product, MC-100 Microfine Slag is a slag-based prod-

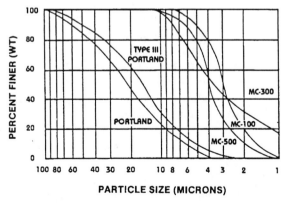

FIG. 5-1. Grain-Size Distribution of Particulate Grout (Clarke 1995)

uct. It has a Blaine fineness of about 12,000 cm²/g with 50% of the particle size less than 3 microns. Fig. 5-1 shows a comparison of grain sizes for Types I, II, and III portland cement, MC-500, MC-300, and MC-100.

Ultrafine cements are more expensive than portland cements. The cost of ultrafine cement can be five times that of portland cement (Moller et al. 1984; Clark 1995). However, a recent development, which should make grouts made with ultrafine cement more economical, is the introduction of the "CEMILL Process" (Bruce et al. 1993). In this process, ultrafine cementitious grouts are produced on site, starting first with grouts made using conventional portland cements. The process involves the reducing of the particle size of the conventional cements to that of ultrafine cement using a special colloidal wet refiner. One key advantage of the CEMILL process is that it makes the use of ultrafine cement grouts independent of local cement markets.

5.3 SAND

Sand is considerably cheaper than cement, so it is used in grout as a filler to reduce the total amount of cement needed. The addition of sand to a grout mix is usually warranted when the water:cement ratio of the grout mix has been reduced to a point of thickening the mix to its minimum pumpable viscosity while the hole still shows no sign of a reduced injection rate. Starting the grouting of the hole with a sanded grout mix may be warranted to fill large open joints, voids, or cavities when they are known to exist.

In addition to the aforementioned economical considerations, sand is also used to reduce the amount of shrinkage experienced by the in-place hardened grout. Both concrete and grout undergo a reduction in volume as they change from the fluid state to the solid state. The amount of the volume change is dependent on the water:cement ratio. The higher the water:cement ratio used, the larger the volume change will be. Fig. 5-2 displays a crude field test showing the reduction in volume of three grouts made with 1:1, 2:1, and 3:1 water:cement ratios. The grout samples in Fig. 5-2 were all cast by filling plastic cups of the same size to the top with freshly mixed grout. They were then allowed to harden overnight. Note the volume change.

Concrete sand meeting the requirements of ASTM C33, "Specification for Concrete Aggregates" can be used for grout. The ASTM C33 gradation limits with respect to sieve size are given in Table 5-1.

As noted in Subsection 4.5, Pumps, sand is abrasive and its addition to a grout mix causes additional wear on pumps and grout distribution systems.

5.4 ADMIXTURES

Admixtures are those ingredients added to grout other than portland cement, water, and aggregates. They are usually added to the mixing water immediately before the other ingredients or are added last, just prior to mixer

MATERIALS 125

FIG. 5-2. Field Test Showing Volume Change of Hardened Grout

discharge. The timing of their addition is based on the type of admixture, the intended results, and the manufacturer's recommendations. Admixtures are used to modify the chemical and physical properties of the grout so as to best suit the project specific requirements.

Many of the same types of commercial admixtures used in concrete mixtures are also employed in grouts. Some admixtures, however, are designed specifically for grout applications. The categories of admixtures used in grouts are normally limited to dispersants, accelerators, and gas-producing agents. Bentonite, as an additive, is discussed later in this chapter. When admixtures are used, the manufacturer's instructions for storage, proportion-

TABLE 5-1. ASTM C33 Gradation Limits

Sieve size	Percent passing by weight
9.5 mm (3/8 in.)	100
No. 4 (4.75 mm)	95–100
No. 8 (2.36 mm)	80–100
No. 16 (1.18 mm)	50–85
No. 30 (600 μm)	25–60
No. 50 (300 μm)	10–30
No. 100 (150 μm)	2–10

ing, mixing and safe handling should be followed carefully. Admixtures can be purchased under many product names. Niche products exist for the country's wide range of regionally different cements, aggregates, and fly ash. These region-specific admixtures are formulated to minimize problems with chemical incompatibility (Rosenbaum 1991).

5.4.1 Dispersants

Dispersants are also called antiflocculents. They are used in grouts to reduce the tendency of the cement particles to agglomerate or flocculate. With a reduced tendency to agglomerate, the grout's ability to penetrate fractures and small openings is enhanced. Dispersants significantly reduce the viscosity, and thus the cohesion, of relatively thick grouts. While there are dispersant admixtures made specifically for this purpose, certain admixtures sold as fluidifiers and superplastizers for concrete also act as dispersants in grouts.

5.4.2 Accelerators

Accelerators are used to shorten the set time for cement-based grouts. Shorter set time can also be achieved by using Type III high-early strength portland cement rather than Types I and II. Calcium chloride ($CaCl_2$) is the material most commonly used as an accelerating admixture. It should conform to the requirements of ASTM D98 and should be sampled and tested in accordance with ASTM D345.

Calcium chloride should be added to the grout mix in solution form as part of the mixing water. If added to the grout in dry form, all of the dry calcium chloride particles may not be completely dissolved during mixing. The use of calcium chloride in the grout mix can cause an increase in drying shrinkage and potential reinforcement corrosion.

The amount of calcium chloride added should be no more than is necessary to produce the results desired for the specific grouting application. An overdose can result in placement problems such as plugged grout lines or premature refusal. Calcium chloride accelerator concentrations of 1–6% by weight of cement are typical. Laboratory and field testing should be performed to optimize the concentrations of accelerator used. Also, concentrations may require adjustments based on seasonal temperature variations at the project site.

Sodium chloride in concentrations of 1.5–5% by weight of cement can be used if calcium chloride is not available, but is less effective (Smith 1987). Other accelerators, such as alkali hydroxides and carbonates, triethanolamine, and high-alumina cements have been used in cementitious grouts (Klein and Polivka 1958). Sodium hydroxide is also a potential accelerator when rapid set is needed. Set times of one hour or less reportedly can be achieved using a 50% solution of sodium hydroxide in a proportion of 2% by weight of cement (Weaver 1991).

5.4.3 Gas-Producing Agents

Gas-producing agents are used to offset the effect of volume reductions, or shrinkage, of portland cement–based grouts as they set and harden, and to reduce bleeding. Finely divided metals, such as zinc, aluminum, and magnesium, added to the mix react with alkalis in the cement to produce hydrogen gas, which induces expansive properties within the grout mixture. However, most gas-producing products on the market are a blend of expanding, fluidizing, and water-reducing agents. These blends produce a slow, controlled expansion prior to the grout hardening. An example is Interplast-N, manufactured by Sika Corporation. In their technical information, Sika specifies a dosage of 1% by weight of cementitious material (portland cement and fly ash, if used). Another product is Flowcable, manufactured by Master Builders Technology (MBS). In their technical information, MBS specifies a dosage of 6% by weight of cementitious material.

5.5 WATER

As a general rule, any water that is drinkable and has no pronounced taste or odor can be used in a grout mix. Some water that is not fit for drinking, however, may be suitable for use in grouts.

The following ASTM designations, C94 "Standard Specification for Ready Mix Concrete," C109 "Test Method for Compressive Strength of Hydraulic Cement Mortars (Using 2 in. or 50 mm Cube Specimens)," and C191 "Test Method for Time of Setting of Hydraulic Cement by Vicat Needle," can be used to aid in determining the suitability of the mixing water.

The grout program designer should select the water acceptance criteria carefully. For example, if the water acceptance criteria are set higher than is technically necessary for the specific project requirements, unjustified additional costs may be added to the project. A situation where this could happen is when a designer copies the water acceptance requirements from an existing reinforced concrete specification into a grouting specification. While most commercial and on-site concrete batch plants can meet or exceed such water-quality requirements, the same water-quality may not be readily available on site for the grouting operation. This same standard may not be necessary for the project's grouting program. However, to meet the unnecessarily high water-quality standards required in the specification, water may have to be transported in, a supply line installed, or a well drilled. Any of these measures will incur additional, possibly unwarranted costs.

In the above example, the upper limits of the chloride concentration in the mixing water for the reinforced concrete may have been established due to the possible adverse effects of chloride ions on the corrosion of reinforcing steel in the concrete. A higher allowable limit of chloride concentration for the grout may be acceptable, however, if it is not going to be in contact with steel reinforcing, embedments, or linings.

5.6 BENTONITE

Bentonite is a colloidal clay from the montmorillonite group that is hydrophilic, or water swelling. Some bentonite can absorb as much as five times its own weight in water (Brady and Clauser 1986). Bentonite is added to cement grout to stabilize the mix by reducing settlement of cement particles. It increases both the viscosity and cohesion of the grout. Bentonite is proportioned to the grout mix as a percentage of cement by weight. The amount of bentonite used usually ranges from 1–4% by weight of cement.

CHAPTER 6

SPECIFICATIONS

There are essentially only two types of specifications used for construction projects. They are referred to as "methods specifications" and "performance specifications."

A methods specification is the one most often used for underground grouting. In a methods specification the contractor is told how to execute the work. The owner is then responsible for how the completed work performs. For example, a methods specification for grouting would dictate such things as the type and quality of cement to use, mix designs, the types and sizes of equipment to employ, acceptable drilling methods, minimum and maximum injection pressures allowed, and refusal criteria. Additionally, the contract drawings would prescribe such other requirements as the diameter, spacing, and depth of grout holes, grout ring geometry and layout, and grouting sequence.

A performance specification, on the other hand, directs how the finished product is to perform. The contractor, not the owner, is responsible to ensure that all of the specified performance criteria are met for the finished product. The contractor is free to use any techniques he chooses to achieve the performance requirements.

Grouting specifications should be written expressly for the project and based on all of the available information. Use of existing specifications from other projects should be discouraged unless each section is carefully reviewed for applicability. Geologic field and laboratory data collected during the site investigation provide the primary basis for the grout program design and specification preparation. The specification writer should, however, search for additional data beyond that produced from the site investigation. Other valuable resources are daily logs, inspection reports, and photographic records from other ongoing and completed projects in the general vicinity of the proposed project. Site visits and interviews with personnel engaged in ongoing projects in the same area are strongly encouraged.

The person assigned the responsibility of writing the grouting specification should be knowledgeable of all aspects of underground grouting. The ideal candidate should have a strong technical background in grouting methods and applications, as well as field experience as an underground grouting engineer or inspector.

One of the things that sets a grouting specification apart from most other specifications is that it must be written to allow maximum flexibility to modify requirements to best suit the actual field conditions encountered at the site. To allow this flexibility, an experienced grouting engineer or inspector, empowered to direct grouting operations, should be assigned in the field during the entire grouting operation. To achieve the best possible grouting results, this person must understand the specification writer's design assumptions and desired results. They must also have the authority to modify the specification requirements to suit the actual field conditions.

6.1 APPLICABLE ASTM AND OTHER STANDARDS

ASTM, the American Concrete Institute (ACI), and the American Petroleum Institute (API) provide laboratory and field testing procedures for grouting materials and for grout mixes. These standards, referred to by number and title, can be written into the specification, and thus become project requirements. Including these standards reduces the time and cost in preparing grout specifications. Also, experienced grouting contractors are familiar with these standards and are comfortable when working with them. Their use can also help to reduce misunderstanding with interruption of the specification.

When standards are specified, they should be applied in their entirety. Attempts to modify the requirements of these standards should be discouraged. If, however, only certain sections of the standards are applicable to a particular grouting project, the specification writer should not hesitate to incorporate only those parts pertinent into the specification.

Table 6-1 contains a listing of the ASTM standards commonly found in grouting specifications. A review of the latest editions of the ASTM standards should be conducted during the specification preparation to ensure the specifying of the latest revision of the standard.

6.2 DEVELOPING GROUTING SPECIFICATIONS

As stated earlier in this chapter and in Chapter 3, underground grouting specifications should be written by an experienced underground grouting engineer, preferably one with field experience. They should also be written

SPECIFICATIONS

TABLE 6-1. ASTM Standards in Grouting Specifications

Standard (1)	Description (2)
ASTM A120	Standard Specification for black and hot-dipped zinc-coated (galvanized) welded and seamless steel pipe for ordinary uses
ASTM C33	Specification for concrete aggregates
ASTM C94	Standard specification for ready mixed concrete
ASTM C109	Test method for compressive strength of hydraulic cement mortars (Using 2 in. or 50 mm cube specimens)
ASTM C117	Standard test method for materials finer than 75 μm (No. 200) sieve in mineral aggregates by washing
ASTM C136	Standard method for sieve analysis of fine and coarse aggregates
ASTM C144	Specification for aggregate for masonry mortar
ASTM C150	Standard specification for portland cement
ASTM C191	Test method for time of setting of hydraulic cement by vicat needle
ASTM C204	Standard test method for fineness of portland cement by air permeability apparatus
ASTM C494	Specification for chemical admixtures for concretes
ASTM C618	Specification for fly ash and raw or calcined natural pozzolan for use as a mineral admixture in portland cement concrete
ASTM C937	Specification for grout fluidifier preplaced-aggregate concrete
ASTM D98	Specification for calcium chloride
ASTM D345	Method of sampling and testing calcium chloride for roads and structural applications
ASTM D3871	Test method for purgeable organic compounds in water using headspace sampling

to allow maximum flexibility to be modified in the field to match the actual field conditions, which helps to yield optimum grouting results.

The author examined several underground grouting specifications written by various federal agencies and private engineering firms and found that there is a prevalent organization and many technical requirements common throughout the industry. These common attributes can serve as a guide or checklist for the specification writer when preparing a grouting specification. A composite list of typical sections found in underground grouting is contained in Table 6-2.

TABLE 6-2. Typical Grouting Specification Sections

PART 1.0 GENERAL
  1.1 Scope of work
  1.2 Reference documents
  1.3 Related work
  1.4 Reference drawings
  1.5 Technical submittals
  1.6 Definitions
  1.7 Measurement and payment
  1.8 Scheduling requirements
  1.9 Work plan requirements

PART 2.0 PRODUCTS
  2.1 Materials
    2.1.1 General
    2.1.2 Cement
    2.1.3 Sand
    2.1.4 Admixtures
    2.1.5 Water
    2.1.6 Bentonite
    2.1.7 Fly ash

PART 3.0 EXECUTION
  3.1 General
  3.2 Material storage
  3.3 Equipment
    3.3.1 Drills
      3.4.1 Percussion
      3.4.2 Rotary
    3.3.2 Mixers
    3.3.3 Agitators
    3.3.4 Water meters
    3.3.5 Pumps
    3.3.6 Pressure gauges
    3.3.7 Packers
    3.3.8 Nipples
    3.3.9 Delivery and distribution systems
  3.4 Drilling
    3.4.1 Minimum and maximum diameter
    3.4.2 Hole alignment
    3.4.3 Hole cleaning
  3.5 Grout mixing and injection
    3.5.1 Mix design
    3.5.2 Materials testing
    3.5.3 Delivery pressure
    3.5.4 Refusal criteria
  3.6 Safety and environmental
    3.6.1 Safety
    3.6.2 Environmental

PART 4.0 QUALITY CONTROL
  4.1 Preconstruction requirements
  4.2 Certifications
  4.3 Records
    4.3.1 Drilling
    4.3.2 Grout mixing and injection
  4.4 Testing

CHAPTER 7

GROUT HOLE LAYOUT

The drilling method, either rotary or percussion, and the type and size of the drills to be used are selected based on such factors as rock type, depth of holes required, working space available, and other factors as described in detail in Chapter 4. This chapter discusses other aspects of drilling such as grout hole spacing, typical hole diameters, hole depths and grout ring spacing for contact, consolidation, and curtain grouting methods.

7.1 GROUT HOLE AND GROUT RING SPACING

The term "hole spacing" refers to the distance between grout hole centers at the hole collar. The hole spacing in a tunnel is the hole-to-hole distance at specific locations (stationing) along the tunnel alignment. Several holes are normally drilled radially at specific stations along the tunnel alignment. This grouping of holes is referred to as a "grout ring." Grout rings are used in contact, consolidation, and curtain grouting. The hole spacing can be given as a measured distance expressed in meters or feet measured at the hole collar on the tunnel perimeter with an angular orientation. More often, however, the hole spacing is given using only the angular orientation in degrees. The intersection of the horizontal and vertical axes of the tunnel is used as the vertex of the angles with a reference of zero degrees as vertically up (Fig. 7-1). The hole spacings for noncircular tunnels, chambers, shafts, and other underground structures are given as the center-to-center distance, horizontally and vertically, between holes at the hole collars. The spacing is expressed in meters or feet. If the hole orientation is other than normal to the face of the rock, then the angular orientation is measured from the hole collar (Fig. 7-2).

7.1.1 Contact Grouting

As discussed in Chapter 2, the purpose of contact grouting is to fill voids between cast-in-place and precast concrete linings and the foundation material. It is also used to fill voids between backfill concrete, being used in

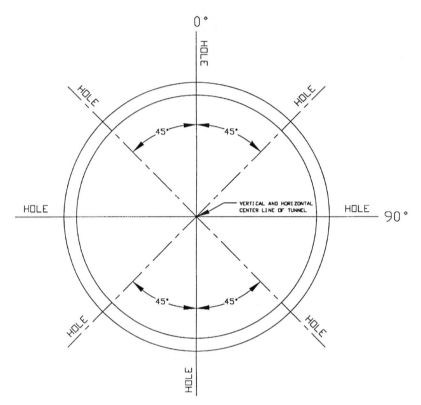

FIG. 7-1. Layout of Eight Hole Consolidation Pattern Using Angular Orientation

conjunction with steel linings, and the foundation material. Determination of the required number and spacing of the holes can be based on the geometry of the structure and standard industry practices for concrete placement. Typical hole and ring spacings are shown in Figs. 7-3, 7-4, and 7-5. These arrangements shown are common for circular tunnels.

Other factors may also affect the number and size of voids and must be taken into consideration. The slump, pumpability, and flowability of the concrete, the workmanship during concrete placement, and the geometry of the steel reinforcement are all consequential to the number and sizes of the voids created. For these reasons, the contact grouting program must remain flexible to allow modifications in the number and spacing of holes to best suit field conditions.

A simple trial-and-error method for locating and determining the size of voids in cast-in-place linings has been successfully used by the author. A jack leg drill is employed to bore a 32–45 mm (1.25–1.75 in) diameter probe hole

GROUT HOLE LAYOUT

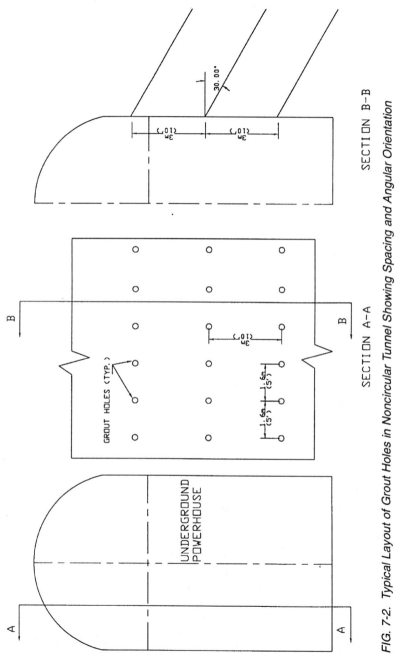

FIG. 7-2. Typical Layout of Grout Holes in Noncircular Tunnel Showing Spacing and Angular Orientation

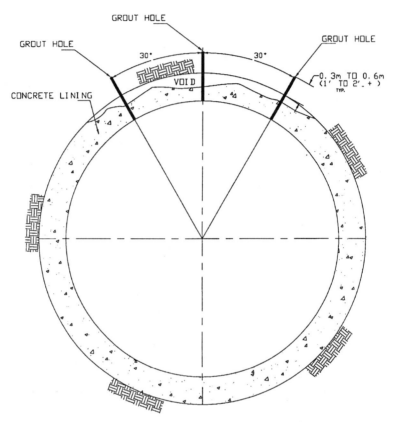

FIG. 7-3. Typical Void and Contact Grout Hole Locations Associated with Cast-in-Place Tunnel Lining

in areas of suspected voids. If a void is found, the probe hole can be used as a contact grout hole; however, additional holes may also be required. If no void is found, the hole is simply filled with a hand-placed (dry-packed) mixture of cement, sand, and water, as opposed to attempting to pressure grouting. This helps save time and money for the overall program.

Concrete liners for underground structures are typically designed 229–305 mm (9–12 in.) thick. At these thicknesses voids can usually be detected visually with a flashlight. The actual lining, however, can be up to a meter or more thick because of overexcavation or overbreak. It thus becomes more difficult to visually detect small voids because the contact grout hole depth increases with the increase of the concrete thickness. To aid detection and measurement of voids, the author has developed and employed a simple tool that can be field fabricated. This tool (Fig. 7-6)

GROUT HOLE LAYOUT

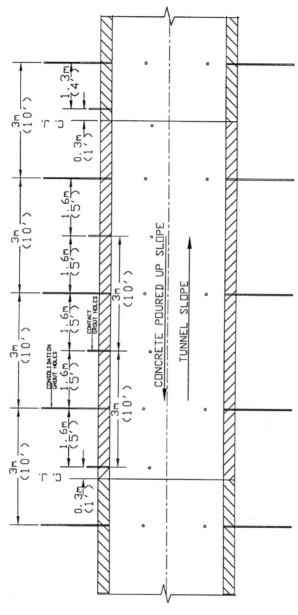

FIG. 7-4. Longitudinal View Illustrating Location of Contact and Consolidation Grout Holes when Tunnel Slope and Cast-in Place Concrete Tunnel Liner Are Placed in Opposite Directions

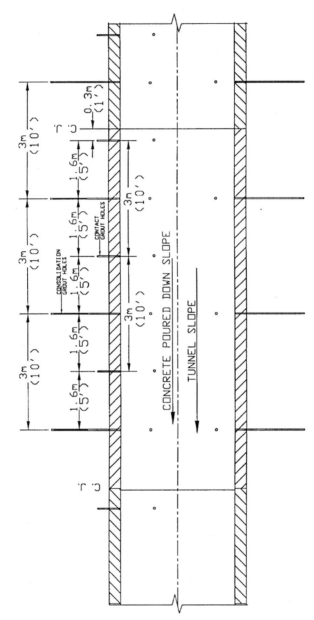

FIG. 7-5. Longitudinal View Illustrating Location of Contact and Consolidation Grout Holes when Tunnel Slope and Cast-in-Place Concrete Tunnel Liner Are Placed in Same Directions

GROUT HOLE LAYOUT

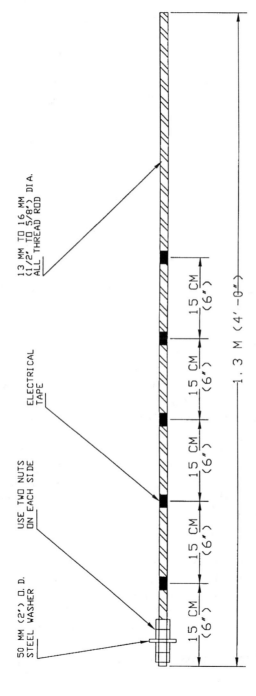

FIG. 7-6. *Field-Fabricated Tool Used to Measure Thickness of Concrete Liner and Size of Void between Concrete and Rock*

provides a measurement of the distance, or void, between the concrete and the rock. The tool can also be used to measure the thickness of the liner concrete actually in place if there are structural concerns about minimum concrete thicknesses. To use the tool, the washer is placed in contact with the wall of the borehole and pushed in slowly. If there is a void, the washer is slipped into it. In addition to identifying a void, this method will give the tool user a measure of the concrete thickness at that location. The washer is further advanced until it hits the borehole in the rock. This distance is the depth of the void. Tape marks, which are used to speed up the taking of measurements in the field, are placed every 150 mm (6 in.) along the rod. By using a ruler to measure between the tape marks, the exact concrete thickness and depth of the void can be determined.

7.1.2 Consolidation Grouting

In Chapter 2, use of consolidation grouting to fill open joints, separated bedding planes, and other defects in the rock up to some specified distance beyond the excavation limits was discussed. This distance is usually taken as a minimum of one tunnel diameter beyond the excavation limits in the case of tunnels.

Data on the geology and structural characteristics of the rock at the project site should be gathered and analyzed during the early stages of the project design. Based on the geologic data, the geometry of the excavated opening, and the excavation method used, (i.e., drill-and-blast or mechanical), the number of consolidation holes required as well as hole depths and hole and ring spacings can be specified.

Figs. 7-7 and 7-8 illustrate typical hole-to-hole spacing for consolidation grout holes for circular tunnels with diameters of 2.5–6 m (8–20 ft). Figs. 7-4 and 7-5 illustrate typical ring-to-ring spacing for consolidation grout holes for circular tunnels. When performing consolidation grouting in underground structures other than circular tunnels, a pattern of 3 m by 3 m (10 ft by 10 ft) is normally used. However, in tunnels and in other structures, a concentration of tightly spaced consolidation holes may be required to allow grouting of geologic structures such as faults, water-bearing zones, and large open-joint systems. When these geologic conditions are known to exist, the possibility of additional grouting requirements should be made clear in the contract documents. However, the length and angular orientations of these holes should be adjusted to best accomplish thorough grouting based on field conditions.

As in a contact grouting program, a consolidation grouting program must remain flexible to allow modification to best suit the actual conditions encountered in the field.

GROUT HOLE LAYOUT

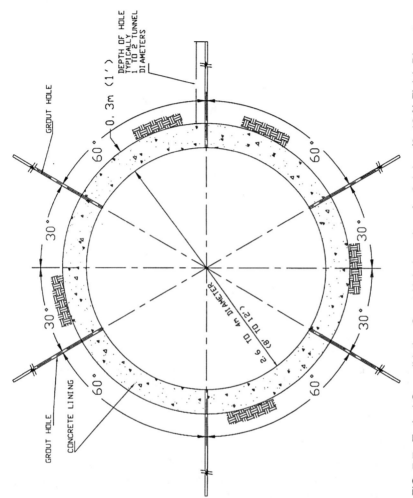

FIG. 7-7. Typical Consolidation Grout Hole Locations for 2.4-4 m (8-12 ft) Finish Diameter Tunnel with Cast-in-Place or Precast Lining

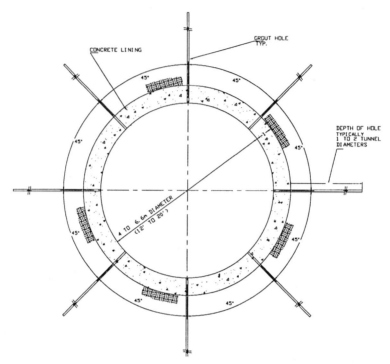

FIG. 7-8. Typical Consolidation Grout Hole Locations for 4–6.6 m (12–20 ft) Finish Diameter Tunnel with Cast-in-Place or Precast Lining

7.1.3 Curtain Grouting

Chapter 2 also detailed curtain grouting, the purpose of which is to reduce or cut off the seepage of water downstream of or beyond the curtain.

The hole spacing considerations for underground grout curtains are the same as these used for dam curtain grouting performed from the surface. The split-spacing method is employed using primary, secondary, tertiary, and quarternary holes (Fig. 7-9). All the holes, drilled radially from the same location, are collectively referred to as a grout curtain ring. A minimum of three rings at 3 m (10 ft) centers is normally used (Fig. 7-10). The horizontal distance between the two outer rings is called the grout "curtain width." The ring that is the furthest downstream is grouted first starting with the primary holes. When all primary holes have been grouted to refusal, the secondary holes are drilled and then grouted. It is important to prohibit the drilling of any of the secondary holes until the primary grouting is complete. The use of tertiary and other higher-order holes may or may not be required based on site-specific geologic conditions and design requirements of the curtain's

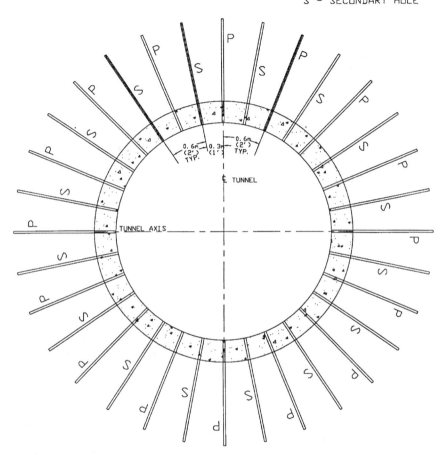

FIG. 7-9. Typical Section of Tunnel Grout Curtain Showing Primary and Secondary Grout Hole Spacing (Section A-A from Figure 7-10) (Tertiary and Quartenary Holes Are Left Out for Drawing Clarity)

permeability. When higher-order holes are needed, they can be offset 0.3–0.6 m (1–2 ft) upstream or downstream of the main ring.

In a three-ring curtain, when the downstream ring is completed, the sequence continues by grouting the ring furthest upstream next with the middle ring grouted last. The same procedure is followed for the upstream ring as with the downstream ring, that is, primary holes are completed before moving onto higher-order holes. The middle ring, which is grouted last, is

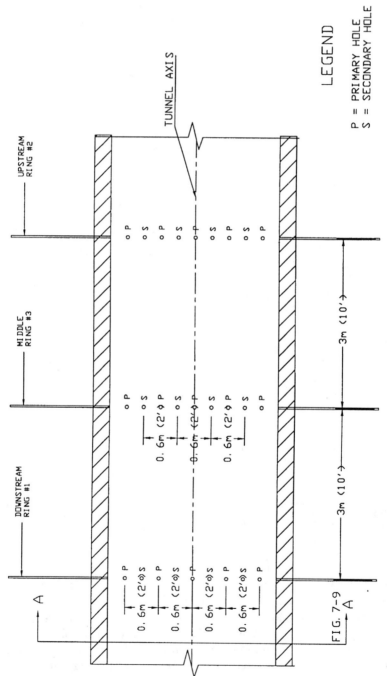

FIG. 7-10. Longitudinal View of Tunnel Grouting Curtain Showing Primary and Secondary Grout Hole Spacing (Tertiary and Quartenary Holes Are Left Out for Drawing Clarity)

considered the "closure ring" and is usually grouted to a higher standard, often utilizing ultrafine cements.

7.2 HOLE DIAMETER AND DEPTH

The range of grout hole diameters typically used underground is between 32 and 63 mm (1.25 and 2.5 in.). The hole diameter chosen by the contractor will generally be governed by the drilling equipment selected, the type and size of packers available, and the diameters of drill holes used for other drilling activities on the project, for example, rock-bolt drilling and blast-hole drilling. For these reasons, the contractor should be allowed to use the widest range of acceptable drill-hole diameters. The range should be specified in the contract documents. By allowing the contractor this flexibility, the overall cost of the grouting program to the owner is reduced.

The depth of grout holes depends on the application. Contact grout holes are usually run the thickness of the concrete liner plus 150–300 mm (6–12 in.) into rock. Consolidation grout holes are usually drilled to a minimum depth of one tunnel diameter in the case of tunnels and are typically 2.4–9.8 m (8–30 ft). Grout curtain holes can range from 9.1 to 30.5 m (30 to 100 ft), but can be much deeper. The hole depth requirements of the curtain are controlled by the geology of the surrounding rock and the operating pressures of the completed structure. Because of the relatively long hole length and large number of holes required for curtain grouting, the possibility of installing the grout curtain from the surface should be carefully considered by the designer. This is usually only possible at portal areas since most pressure tunnels, reservoirs, and underground powerhouses are generally located too deep to make surface drilling practical.

CHAPTER 8

GROUT PLACEMENT OPERATION

The procedures for drilling, proportioning and mixing, and injection used in underground grouting are basically the same as those employed when grouting from the surface, except that access and workspace availability are usually limited. The negative effects of these limitations can be reduced by good advanced planning of the overall grouting operations, careful selection of equipment, and assigning an adequately sized and experienced work crew.

8.1 PROPORTIONING AND MIXING

The proportioning and mixing of the grout should be performed as close as possible to the point of grout injection. However, if because of workspace limitations, coordination with other work activities, or anticipation of very large grout takes, remote or aboveground proportioning and mixing is often necessary.

8.1.1 Water:Cement Ratios

In underground applications water:cement ratios are expressed in terms of volume or in terms of weight. The volumetric ratio is preferred on smaller grouting projects and is particularly easy to use when bag cement is used. A bag of portland cement has a loose dry volume of approximately 28.3 L (1 ft³) and weighs 42.6 kg (94 lb). However, when a bag of cement is added to water it occupies approximately 14.2 L (0.5 ft³). For example, if one bag of cement is added to 28.3 L (1 ft³) of water, a water:cement ratio of 1:1, the resulting mixture would occupy a volume of approximately 42.4 L (1.5 ft³). Further, if one bag of cement is added to 56.6 L (2 ft³) of water, a water:cement ratio of 2:1, the resulting mixture would occupy a volume of approximately 70.7 L (2.5 ft³). When the volumetric method of proportioning is used, the water meter should be calibrated in liters or cubic feet. The use of water meters calibrated in gallons should be avoided.

The quantity of water required for hydration of cement is approximately 0.45:1 by volume, or 0.3:1 by weight (Weaver 1991). Water added beyond these ratios is used to make the grout mixture pumpable and to allow maximum penetration of joints, cracks, and other openings in the foundation material.

In the past, standard practice has been to use very thin grouts with water:cement ratios of 5:1 and greater. In theory these thin mixes would achieve maximum penetration of the foundation materials. Research, however, has shown that grout placements using water:cement ratios of 5:1 and higher (i.e., thinner grout) tend to leach, whereas placements using 3:1 or lower ratios (i.e., thicker grout) have been used successfully and produce a higher quality grout once in place (Houlsby 1982, 1985). Therefore, the thinnest grouts, used at the start of grouting for each hole during consolidation and curtain grouting should have a water:cement ratio of no greater than 3:1. However, when contact grouting, a grout with a water:cement ratio no greater than 2:1 should be used at the start of grouting.

When proportioning and mixing a batch of grout, the total quantity of mixing water should be added to the mixer first, before any cement or other ingredients. The cement should be added slowly, passing it through a metal screening or hardware cloth. This is done to sieve out hardened lumps of cement and pieces of the cement bag that could plug the grout line or cause premature hole refusal.

A colloidal mixer, as opposed to a paddle mixer, should be used for mixing of consolidation and curtain grouts. When using a colloidal mixer a minimum mixing time of 15 sec is adequate for complete mixing if the grout consists of only cement and water (neat grout). When bentonite is used, a minimum mixing time of 1 min should be used. However, the mixing time should not exceed 1 min, because the heat generated during mixing could cause the cement to hydrate prematurely and may prevent the grout from setting in place after injection (Water Resources Commission of New South Wales, 1981).

8.1.2 Sand

The use of sanded grouts underground is usually limited to contact grouting. It is also used to grout open joints or when large extensive voids and caves are encountered in the foundation material during consolidation and curtain grouting.

When sand is required as a filler it is proportioned in terms of relative volume with respect to water and cement, if batching is done by the volumetric method. For example, a 1:1:0.5 mix would contain 28.8 L (1 ft^3) of water, 28.8 L (1 ft^3) of cement, and 14.4 L (0.5 ft^3) of sand. A 1:1:0.5 to 1:1:1 ratio mix is commonly used for contact grouting. A box made of wood or steel with a volume of 14.2 L (0.5 ft^3) can be used to measure the sand. When large-capacity batching operations are employed, sand, and other ingredients are added by weight. It is highly recommended to use bentonite with sanded mixes to help hold the sand in suspension. The water, cement and bentonite should be thoroughly mixed before any sand is added to the mix. When sand is used, the grout is usually discharged first from the colloidal mixer into a paddle mixer or another colloidal mixer where sand is then added.

8.1.3 Admixtures

Admixtures are supplied in either a dry powdered form or as a liquid. Some admixtures are added to the mix water before the cement is added, and other admixtures are added to the neat cement grout after it has been thoroughly mixed. The specific admixture manufacturer's instructions for proportioning and mixing should always be followed. A more detailed discussion of admixtures is found in Chapter 5.

8.1.4 Bentonite

When bentonite is used it should be added slowly to the mix water and thoroughly mixed before any cement is added. Ideally, when large quantities of bentonite are expected to be needed, the bentonite is prehydrated before it comes in contact with the cement slurry. To achieve prehydration requires separate mixing and storage of the bentonite and water mixture in a large agitator tank. The minimum prehydration time based on studies at Tucurui Dam is 8–12 hr (Weaver 1991). Since the amount of bentonite used in a mix is relatively small, 2–8% of the weight of cement, it can be premeasured and bagged in plastic sacks weighing 0.9–3.6 kg (2–8 lb) when bagged cement is used.

8.2 DELIVERY PRESSURE

In the United States, a long standing rule of thumb for calculating the maximum allowable grout injection pressure when grouting rock has been 1 lb/in.^{2} per foot of packer depth in rock plus half the depth of overburden. This rule is based on the weight of rock and overburden directly above the grout hole. While this approach to determining maximum injection pressures may be applicable when grouting in poor or unknown geological conditions at shallow depth, 10 m (30 ft) or less, it is felt to be too conservative for most other rock grouting (U.S. Army Corps of Engineers 1984). To reinforce the philosophy of using higher injection pressures, the European rule of thumb for safe grouting pressures is 1 kg/cm^2 per meter of packer depth. This is about four times the U.S. rule of thumb (Weaver 1991).

In their manual *Grouting Technology*, the U.S. Army Corps of Engineers states that other factors affecting the maximum safe grouting pressure include rock strength, orientation of rock discontinuities, consistency of the grout, tightness of the hole, geology, and hydrologic conditions. In the manual they present a rough guide for grouting pressures, which is presented in Fig. 8-1. As shown in the figure, safe grouting pressures can range from 1 to 4 plus psi per foot of packer depth. Since injection pressure is one of the most important criteria for maximizing the radius of penetration of the grout, the maximum safe injection pressure should be applied (Lombardi 1985; Weaver 1991).

The packer depth when grouting from the surface is the vertical distance from the packer to the closest point on the ground surface. In underground grouting however, this distance must be measured from the bottom of the grout hole to the closest point on the ground surface or other underground structures for holes drilled in an upward direction from the horizontal (see Fig. 8-2).

Another important fact to remember when grouting holes drilled upward is that the static head pressure of the column of grout must be overcome before any positive grouting pressure is applied to the hole. Therefore, the pressure caused by the weight of the column of grout must be added to the gauge pressure applied at the hole collar. This is the exact opposite of what is done when grouting a vertical downward hole, as in the case with surface grouting. When grouting upward, the static pressure to be overcome is determined by multiplying the vertical height of the grout column by the density of the grout, which gives the pressure of the grout column acting downward. And the density of the grout is based on its water:cement ratio (see Table 8-1).

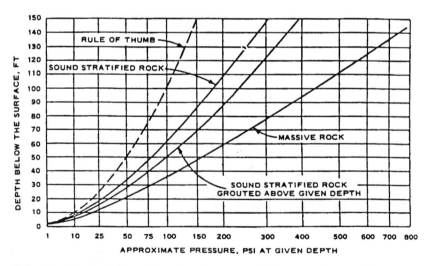

FIG. 8-1. Rough Guide of Grouting Pressures (U.S. Army Corps of Engineers 1984)

GROUT PLACEMENT OPERATION

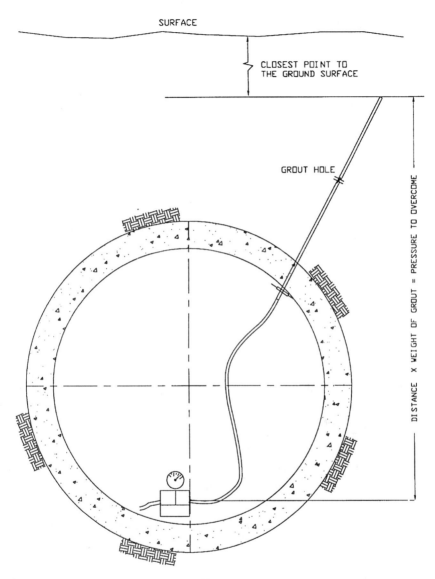

FIG. 8-2. Section Showing Calculation of Allowable Pressure and Static Head of Grout Column

TABLE 8-1. Correction Factor for Pressures of Grout Column Based on Water: Cement Ratio

Water:cement volume (1)	Correction Factors	
	SI (kPa/m) (2)	English (psi/ft) (3)
0.75	17.9	0.789
1.00	16.7	0.736
1.50	14.9	0.660
2.00	13.9	0.614
2.50	13.2	0.584
3.00	12.7	0.562
4.00	12.2	0.541
5.00	11.7	0.515
6.00	10.8	0.476

The following is an example of the calculation for determining the static head pressure given these conditions:

- A vertical 3.3 m (11 ft) deep consolidation hole in rock at the crown of a tunnel
- Use of a 1:1 grout mix
- The bottom of the hole at 15.2 m (50 ft) from the closest point on the ground surface
- The header located 0.3 m (1 ft) off invert of a 3.6 m (12 ft) inside diameter concrete lined tunnel (see Fig. 8-3)

Calculate the gauge pressure required using 552 kPa (80 psi) allowable injection pressure (reference Fig. 8.1, for sound, stratified rock).

To calculate static head pressure use the formula

[(Depth of hole) + (Thickness of concrete liner) + (Distance from crown to header)] × (Density of mix) = Static head pressure

(3.3 m + 0.3 m + 3.3 m) × 16.7 kPa/m = 115 kPa

or

(11 ft + 1 ft + 11 ft) × 0.736 psi/ft = 16.9 psi

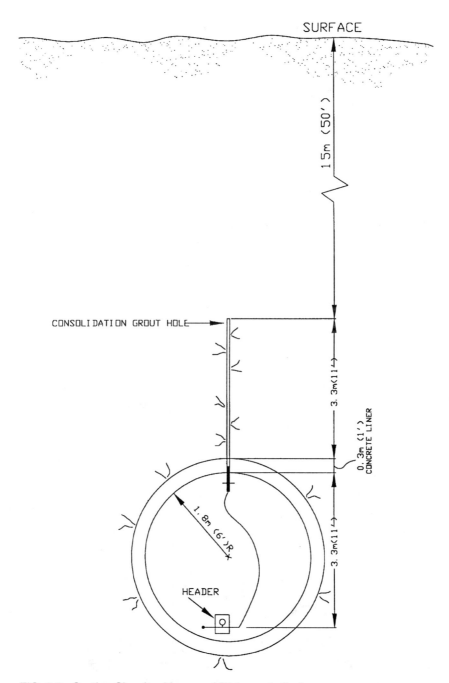

FIG. 8-3. Section Showing Measured Distance to Surface

To calculate total gauge pressure required use the formula

(Static head pressure) + (Allowable injection pressure) = gauge pressure

115 kPa + 552 kPa = 667 kPa

or

16.9 psi + 80.0 psi = 96.9 psig

When grouting underground near a portal, shaft, underground opening, or other underground structure, the horizontal distance from the grout hole to the open face of the opening or structure may be the shortest distance. Thus, these distances, not the ground surface, become the governing factors for calculating the maximum allowable injection pressure (see Fig. 8-4).

The maximum allowable injection pressure for performing contact grouting of shaft and tunnel linings is approximately 207 kPa (30 psi). This is also

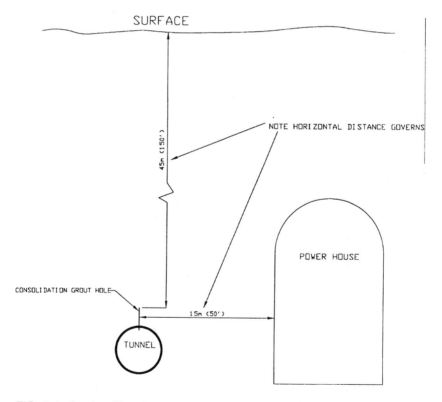

FIG. 8-4. Section Showing Horizontal and Vertical Distances Measured from Consolidation Grout Hole to Nearest Structure

a safe maximum pressure when performing backpack grouting of steel liners and transition sections. However, a check calculation to establish the maximum allowable injection pressure should be performed when grouting steel linings. The steel can easily become distorted from overstressing.

8.3 REFUSAL CRITERIA

Refusal is the point when the grout injection is stopped on a specific hole. The refusal criteria used can have a potentially important effect on whether or not a significant reduction in permeability can be achieved in rock containing relatively fine fractures (Weaver 1991). The refusal criteria should be given in the specifications for each grouting method to be used on the project. Refusal criteria is given as a minimum amount of grout injected into the hole per a specified time period at the maximum allowable pressure. An example would be as follows: less than 0.3 m^3 or 28.31 L (1 ft^3) in 10 min at the maximum pressure. The refusal criteria requirements should increase in tightness from contact through consolidation to curtain grouting.

Suggested refusal criteria for three types of grouting are

- Contact grouting—less than 0.03 m^3 or 28.3 L (1 ft^3) in 10 min at maximum pressure
- Consolidation grouting—less than 0.003 m^3 or 2.38 L (0.1 ft^3) in 10 min at maximum pressure
- Curtain grouting—no take, as indicated by no change in dipstick reading at the agitator in a 15 min period (Water Resources Commission of New South Wales 1981)

Specifications may also be written so as to require the stopping of grouting on a specific hole when a stipulated quantity of grout or cement has been injected into the hole. An example of such instruction is "grouting of a hole shall be stopped when 150 sacks of cement have been injected into the hole."

8.4 CREW SIZE AND ORGANIZATION

A typical underground grout placement crew has five personnel: the mixer/pump operator; a person to install packers and hook up and move the delivery lines from hole to hole; a header operator; a material delivery equipment operator; and a supervisor. A typical underground drilling crew consists of one driller per drill plus one driller helper per two drills. Placement and drilling crew sizes may vary depending on local area labor practices and collective bargaining agreements.

8.5 PRODUCTION

The production rates for drilling of grout holes and the mixing and injection of grout varies greatly. Drilling rates, usually expressed in feet or meters per hour or per shift, are dependent on factors such as

- Rock type, hardness, and abrasiveness
- Structural geology
- Hole diameter, length, and orientation
- Type and size of drilling equipment
- Working space available
- Number of holes located at a single drilling location or grout ring
- Distances between drilling locations
- Work crew experience
- Local labor practices

The time required to drill a hole, in addition to the actual drilling time, which are both independent of hole depth, comprise the move-in, setup of the drill, and orienting of the drilling equipment on the hole. The length of individual pieces of drill rod is dependent on the drilling equipment type and the size of the working area within the underground opening. The shorter the length of a drill rod, the greater the number of drill rods required for a given length of hole.

The time required to mix and inject grout is dependent on factors such as

- Capacity of the mixer
- Working space available
- Number of holes located at a single grout location or grout ring
- Number of different mixes (water:cement ratios) used for each hole;
- Total amount of grout injected into the hole
- Distance between grouting locations
- Work crew experience
- Local labor practices

The following production rates are offered as a guide to allow preliminary scheduling of the grouting portion of a project. Once drilling and grouting are under way, the schedule can be updated and adjusted to reflect actual rates and site-specific conditions. Drilling rates require

- 1–2 min—Setting the drill rig at the proper orientation, including moving from one hole to the next at a single grouting location
- 1–3 min/m (0.25–1 min/ft)—Drilling rate, which is dependent on hole diameter, the rock type and hardness, structural geology, and the size and type of equipment
- 0.5–1 min/rod—Adding and removing drill rods, dependent on the hole depth, the drilling equipment type, and drill rod length
- 15–30 min—Moving the drilling equipment from one location

GROUT PLACEMENT OPERATION 157

to the next assuming 3–6 m (10–20 ft) between locations; also dependent on the size and type of equipment

Mixing and injection requires

- 15 min—Set up at the beginning of shift, including mixing the first batch of grout
- 2 min/packer—Setting packer in hole, assuming the packer is located at the hole collar; can be performed by other crew members at the same time as the rest of the crew is setting up the mixture
- 2 min/hole—Disconnect delivery line from packer and reconnect line to next packer
- 15 min/hole (minimum)—Actual grout injection (a quite variable rate)
- 30 min—Breakdown and cleanup at end of shift or when moving from one location to another
- 30 min—Moving mixing equipment and materials from one location to the next assuming 15–30 m (50–100 ft) between locations

8.6 SAFETY AND ENVIRONMENTAL ISSUES

Worker safety should be a high priority on all construction and mining projects. Working underground in often restricted work areas, working with cements and chemicals, working under less than ideal lighting conditions, the need for ventilation, and dealing with high noise levels increase the need for careful attention to safety.

Environmental laws and regulations are becoming more stringent and widespread. In addition to the federal government, many state and local government agencies are adopting their own laws and regulations governing underground work. Due to the diversity of underground projects they can take place in wilderness areas or in the middle of a large urban areas; thus quite different environmental issues so apply. Compliance with various regulations is time-consuming and expensive. Issues such as the time required to obtain environmental permits must be factored into the project schedule. Real estate requirements as well as the time and cost to construct facilities such as settlement ponds need to be considered. Environmental issues need to be taken very seriously when planning, scheduling, and costing a project.

8.6.1 Personnel Safety

Safety of personnel working in the underground construction industry is governed by United States Department of Labor Occupational Safety and

Health Administration (OSHA). Safety of personnel working in the underground mining industry is governed by the Mine Safety and Health Administration. Copies of these regulations are available from the respective federal agencies. In some areas, state and local safety regulations, which are sometimes more restrictive than federal regulations, may also govern.

While all laws must be followed, several areas of personnel safety are particularly important in underground work and deserve special attention. These include issues such as

- Restricted working space
- Chemical burns from cement and admixtures
- Adequate lighting
- Proper ventilation
- High noise levels
- Poor conditions for verbal communication
- Dust control
- Protection from moving and rotating equipment
- Access and egress from work areas to the surface

The problem of restricted working space can be mitigated by selecting the pieces of equipment that allow the maximum clearances between the equipment and the structure. Work injuries can be reduced by ensuring that protective guards are installed around moving and rotating equipment parts. General safety and housekeeping can be improved by keeping walkways clear of grout hoses, equipment, and other tripping hazards as well as by keeping the work area clean of empty paper cement bags, spilled grout, and other materials.

Chemical burns are a problem whenever workers are exposed to portland cement–based construction materials. In addition to grout, workers exposed to concrete and shotcrete experience similar hazards. Proper personal protective clothing is the best defense against chemical burns. As a minimum, workers should be required to wear long-sleeve shirts, rubber gloves, and safety glasses. Additionally, every effort must be made to limit workers' contact with cement and grout. This can be accomplished by maximizing the use of machinery to transport, measure, mix, and inject the grout. Material safety data sheets supplied by the suppliers of chemicals and admixtures must be kept on site and made readily available on all shifts. These data sheets list the chemicals in the admixtures and tell how to apply first aid after personnel exposure.

There are minimum lighting requirements, given in foot-candles, found in the regulations. The regulations also specify other technical requirements of underground lighting and power-supply systems. Every effort should be made to increase lighting well beyond the minimum requirements. This will not only increase safety but will help increase productivity. Underground ventilation requirements are given in the regulations as a minimum air velocity within the underground workings and the required chemical makeup of the air in the work area. The regulations also specify other technical requirements for fans, ventilation ductwork, and the power supply for ventilation systems. Additionally, diesel engine equipment used underground requires the installation of a scrubber in the engine's exhaust system. Gasoline-powered equipment is not allowed underground.

High noise levels are common on construction and mining projects. Underground, however, noise generated by equipment operations is particularly troublesome because sound waves reverberate off the surfaces of the underground structures. Some equipment can be equipped with mufflers in their exhaust systems to help reduce noise levels. Nonetheless, ear plugs or ear mufflers should be worn by all workers assigned to grouting operations.

Airborne particulates are also a health and safety issue in the underground work environment. The main sources of dust during grouting operations are from the handling of cement and drilling. Cement dust is hard to control, so dust masks should be worn by workers handling cement. Dust from drilling can be greatly reduced by using water in the drilling operation. There will, however, always be some dust and atomized oils from air-operated equipment in the area around the grouting operation; therefore, dust masks or respirators should be worn by all members of the grout crew.

Access and egress requirements for underground work are also regulated. Main points of underground access and egress are normally established as part of the larger overall project. Grouting operations, however, sometimes take place in remote underground locations that may be a considerable distance from the main construction or mining operations. It is therefore necessary for a responsible person on the surface to know how many people are working in the grout crew and their exact location underground at all times. Also, the entire grout crew must know the escape route and evacuation procedures to use in case of an emergency.

8.6.2 Waste Disposal

Waste is generated by grouting operations as drill cuttings and as unused grout and wash-water. Drill cuttings are usually not considered an environmentally hazardous material and can be removed along with the other excavated materials. However, unused grout, as well as wash-water, used to clean out grout lines and equipment, must be handled specially. The first step is to transport unused grout materials from underground to the surface. The unused grout should be placed or pumped into a plastic-lined wooden or steel box, in which it is allowed to harden. The hardened grout is then removed from the box and disposed of with other excavated materials or transported to a landfill that accepts concrete. Removal of wash water poses several additional steps. The wash-water is generated in the general area of the grouting operations from the flushing of the grout lines and the washing out of mixers, agitators, and pumps. The grout-laden wash-water will flow downhill to a point were it is collected in a sump and pumped to the surface. The flow path, sump pumps, and discharge lines should be flushed with water continuously. If not flushed properly, the flow path, for example, the tunnel invert, will develop a buildup of grout that requires chipping to remove it. The grout-laden wash-water also has the potential to foul pumps and wastewater discharge lines; therefore, continuous flushing is recommended. Once the wash-water reaches the surface, it has to be settled to remove suspended particles before it is discharged into storm or sanitary wastewater systems. A recently enacted law, the National Pollutant Discharge Elimination System (NPDES), was enacted under the General Permit for Storm Water Discharges from Construction Sites, as published in the *Federal Register* in September 1992. This law governs water generated by construction sites that is subsequently discharged off the job site.

CHAPTER 9

FIELD QUALITY CONTROL

A good quality control program helps ensure that technical requirements of the grouting are met and also adds to the overall success of a grouting project. The quality control program for underground grouting starts with the design. The design must be adequate to meet the project engineering requirements while remaining reasonable and sufficiently flexible from a constructability standpoint. This can be best accomplished by engaging an experienced grouting engineer during the early stages of design. This individual will work closely with the design team to establish practical and achievable goals for both the quality control program and the grouting operation as a whole.

One aspect of the quality control program is to write quality into the specifications. This is accomplished by requiring equipment, methods, and materials that are commonly used, have a good record of success, and are readily available. This is not to say that a state-of-the-art or an innovative grouting program should not be considered, but only that the size, location, and sophistication of the project should be considered when developing the overall grouting program.

Another factor contributing to a good quality control program is in the implementation of the program in the field. Again, having an experienced grouting engineer or inspector assigned to the field full time during the entire grouting operation is an important element for the success of the program.

The field quality control effort is made up of several components, beginning with the preconstruction planning and the development of a preconstruction checklist, and then obtaining and verifying material and equipment certifications, developing and maintaining drilling and grouting records, and performing production testing.

In addition to helping maximize the technical results of the grouting, a well-planned and executed quality control program can also help control costs and maintain the schedule for the project. For example, during the preconstruction planning and development of the preconstruction checklist, equipment limitations and other logistical concerns may be discovered, thus allowing corrective action to be taken before field operations begin. Additionally, thorough record keeping can help eliminate pay-quantity disputes between the grouting contractor and the owner or prime contractor. The records can also be used to help analyze production rates, thus allowing construction schedule updating and adjusting when necessary.

Quality control records and test results are also used as a primary data source to evaluate grout program performance during the start-up and early operation stages of a project. Examples of this are the evaluation of excessive water loss during the filling of a water conveyance system, such as a pressure tunnel, or the unexpected or unacceptable rise in the groundwater level in the immediate project area during the watering up of the system. Grouting records were used to help determine the best course of action to reduce the amount of leakage at the Helms Pumped Storage Project in the Sierra Nevada mountains of California (Moller et al. 1984).

9.1 PRECONSTRUCTION CHECKLIST

A preconstruction checklist is an itemized list developed by reviewing the requirements of the project specifications and the contract drawings. These requirements are made up of paperwork submittals such as resumes of the contractor's key personnel, mill test reports, results of the mix water chemical analysis, equipment calibration records, and the contractor's construction schedules and work plans. Additional requirements may involve field verification of equipment sizes and types, proper storage of materials, such as cement, as well as safety and environmental compliance. A sample preconstruction checklist is shown in Fig. 9-1.

9.2 CERTIFICATION AND TEST REPORTS

A dictionary definition of "certification" is "a written statement attesting to some fact." The more common certifications pertaining to grouting are written statements by materials suppliers stating that their product meets a specific specification-referenced standard, such as an ASTM, or other requirements stipulated in the specifications.

Certification may also be required for laboratory test results performed on grouting materials, such as a sieve analysis of sand. Certifications should be required for calibration of equipment like water meters and pressure gauges.

9.3 INSPECTION RECORDS

Inspection reports produced by the grouting engineer or inspector are the primary documents used to record all work activities associated with the grouting operation. In addition to documenting adherence to contract documents, these reports can be used to verify pay quantities. They also serve as a critical database for helping analyze engineering and operational evaluations that might be needed during the start-up, early operations, or life of the facility.

FIELD QUALITY CONTROL

PRECONSTRUCTION CHECKLIST

DESCRIPTION	DATE RECEIVED	DATE APPROVED
• Resumes of key personnel.		
• Mill test reports for cement.		
• Mill test reports for bentonite.		
• Sieve analysis/gradation curve for sand.		
• Manufacturer's data on admixtures.		
• Results of chemical analysis of mix water.		
• Grout mixer manufacturer's spec.*		
• Grout agitator manufacturer's spec.*		
• Grout pump manufacturer's spec.*		
• Pressure gauge manufacturer's spec.*		
• Packer manufacturer's spec./shop drawing.*		
• Contractor's const. schedule and work plan.		
• Waste disposal plan.		
• Contractor's quality control program.		
• Contractor's safety program.		
• Contractor's drilling and grouting reports.		
*(less commonly required)		

FIG. 9-1. Preconstruction Checklist

Two separate sets of reports should be kept: one to record the drilling activities and the other to record the grout mixing and injection operation. In addition to these project-standardized inspection reports, each inspector should be required to maintain a field grouting diary. The diary is kept in a standard hard-covered engineer's field book. The diary is used to record all events not specifically covered on the inspection report or to expand and explain unusual data shown on the report. It is important that the inspector record all events, no matter how trivial they may appear to be at the time. If information is not recorded contemporaneously, it is lost forever. It is better to have the information and never use it than to not have the information at all.

The diary is passed from one inspector to the next on multiple-shift operations. Each inspector should date and sign for their individual shift entries. An entry should be made for each shift even if no unusual events occurred. A simple statement of "no unusual events" with a date and signature is sufficient. A copy of the entries for each shift should be photocopied and turned in at the end of every shift along with the standardized reports. This information should be carefully reviewed and analyzed daily, as a minimum, by the person(s) in charge of the grouting program. Based on the data, modifications, when necessary, can be made to the grouting program in a timely manner.

9.3.1 Drilling Reports

As a minimum, drilling inspection reports should contain all of the technical attributes that are required in the contract documents, such as hole diameter, depth, and spacing. A separate drilling report is kept for each location or grouping of grout holes. The locations should be clearly identified using job-standardized stationing and hole identification and orientations. Also, holes drilled for different types of grouting, for example, contact versus consolidation, are recorded separately. A sample drilling inspection report is shown in Fig. 9-2.

9.3.2 Grouting Reports

Grouting inspection reports contain, as a minimum, all of the technical attributes that are required in the contract documents, such as the mix used, time of batching, injection pressure, and quantity of grout injected into each hole. A separate grouting inspection report is kept for each location or grouping of grout holes. Also, holes for different types of grouting, for example, contact versus consolidation, are recorded separately. A sample grouting inspection report is shown in Fig. 9-3.

9.4 TESTING

The same types of testing are performed for underground grouting as are performed for surface grouting. Preconstruction testing is performed in the laboratory and field testing is performed as the grouting operation is in progress. There may also be laboratory testing, such as grout cube compressive testing, required as the work progresses.

9.4.1 Laboratory Testing

Laboratory testing is used to evaluate the proposed grouting materials as well as chemical and physical properties of the grout mixture. Two common material tests are a chemical analysis of the mixing water (ASTM D3871 *Testing Method for Purgeable Organic Compounds in Water Using Headspace Sampling*) and sieve analysis for sand (ASTM C33 *Specification for Concrete Aggregates*). Additional tests may be required, for example, chemical analysis and gradation of the cement, necessitated by availability and quality of the local cement supply.

FIELD QUALITY CONTROL

Continental Drilling U.S.

Date _____ Location _____ Shift _____
Job # _____ Hole # _____ Temp: _____
Drill _____ Angle _____ High _____ Low _____
Pump _____ Contractor _____ Precip. _____

TIME		Activity	Hole Size	DEPTH		Total	Bit#	Remarks
Start	Stop			Start	Stop			

MATERIALS	
Type	Amount

Comments: _____
_____
_____
_____
_____

Driller Signature _____

Description	Name	Union Local	Pay Rate	Subsistance	Regular Hours	Overtime Hours	Remarks
Driller							
Helper							
Helper							
Water driver							
Foreman							

FIG. 9-2. Sample Grouting Report

CONTINENTAL DRILLING COMPANY

DAILY GROUT REPORT

PROJECT	DATE:	SHEET ______ OF ______ SHEETS
CONTRACTOR:	SHIFT LOCATION:	FIELD BOOK NO.: ______ PAGE: ______

STATION	TIME		PAY CONN.	ZONE	STAGE	DEPTH		W/C RATIO	HOLE PRESS (PSI)	GROUT PLACED (C.F.)	CEMENT (SACKS) (C.F.)	SAND (C.F.)	REMARKS
	START	STOP				FROM	TO						

BID ITEM		SIGNATURES
(C.F.) CEMENT MIXED	SKS.	CONTRACTOR:
(C.F.) CEMENT PLACED ______ SKS.	SKS.	
(C.F.) CEMENT WASTED	SKS.	INSPECTOR:
FLUIDIFIER IN GROUT	LBS.	
SAND USED	CU. FT.	DATE SIGNED:
GROUT HOLE CONNECTIONS	EA.	

CMV—2

FIG. 9-3. Sample Drilling Report

Laboratory testing of the grout mixture is also used to evaluate properties such as set time, cohesion, and compatibility. Set-time test data is used to avoid flushing out of "green" grout as a result of premature drilling of nearby holes. The most commonly used test is performed using the Vicat needle apparatus (Weaver 1991). The procedure used is a modification of ASTM C 191-82, *Time of Setting of Hydraulic Cement by Vicat Needle*. There is no ASTM standard for measurement of cohesion of grout. However, a description of a testing procedure is given by Weaver (1991). Compatibility testing is conducted to evaluate items such as how the site's geologic materials and groundwater will react chemically with the grout; the mineralogy of the mixture's sand and how it reacts with the cement; and the behavior of various proposed admixtures with the other grouting materials.

9.4.2 Field Testing

Field testing consists of checking the quality of the grout in its fluid state just prior to injection and the casting of compressive strength specimens for laboratory testing of cured samples.

The quality of grout in its fluid state is tested for density, viscosity, and sedimentation or bleeding. The American Petroleum Institute has established testing procedures for drilling fluids that are applicable for geotechnical grouting. Table 9-1 shows suggestions for the ranges of test results for bentonite-cement grouts (Weaver 1991).

Specific gravity testing is used to check proper mix proportioning. For example, an excessive specific gravity may be the result of an error in adding cement where an extra bag of cement may have been added. Conversely, a low specific gravity may indicate that a bag of cement was left out of the mix or that too much water was added. The adding of too much water may be the result of human error or a problem with the water-measuring equipment.

The Marsh viscosity test is used to determine the viscosity of grouts with water:cement ratios of 1.5:1 to 3:1 and greater. For thicker grouts the ASTM Flow Cone Test is used. The test results are measured in time with results given in seconds.

Sedimentation testing is performed periodically during the injection operation for each mix or change in the water:cement ratio. The sedimentation

TABLE 9-1. Ranges of test results for bentonite-cement ratios

w:c ratio (1)	Specific gravity (2)	Marsh viscosity (3)	Sedimentation (4)
3:1	78.7–81.9	26.5–28.5	15–30%
2:1	85.7–89.3	32.0–36.0	4–6%
1.5:1	91.9–95.7	45.0–50.0	2–3%

rate of the grout mixture is determined by pouring 1,000 mL of grout into a standard 1,000 mL laboratory cylinder graduated in milliliters. Then, after 2 hr, measurements of the water level and the solids level in the cylinder are taken and recorded. The sedimentation rate, expressed in percent, is obtained by dividing the depth of supernatant water by the total depth of water and solids and multiplying by 100.

Compressive strength testing is performed in the laboratory on specimens of grout taken in the field during production grouting. A set of three, 50 × 50 × 50 mm (2 × 2 × 2 in.) grout cubes is the most common type of specimen used. The cubes are cast, cured, and tested per ASTM C 109 *Test Method for Compressive Strength of Hydraulic Cement Mortars* (Using 2 in. or 50 mm cube specimens). However, rather than cubes, a smaller-size version, usually 50 × 300 mm (2 × 6 in.), of the standard concrete cylinder is sometimes used. If cylinders are used, a correction factor based on the aspect ratio of the smaller cylinder to the concrete cylinder must be used to obtain a true compressive strength. The requirement for the compressive testing of grout is not common for geotechnical grouting applications. However, it is more common for structural grouting such as contact and backfill grouting.

Field testing can also be performed on the in situ materials to be grouted to check the materials' groutability using several proposed mix designs. This testing is done under controlled conditions, usually at the actual job site, utilizing the actual equipment and methods that will be employed for the production grouting.

CHAPTER 10

CHEMICAL GROUTING[a]

Chemical grouting is used in underground construction primarily on soft-ground shafts and tunnels in an effort to control cohesionless or "running" ground conditions by modifying the soil. Chemical grouting involves the injection of a two-part grouting system usually using sodium silicate–based grout, which is injected into the soil mass for the purpose of increasing the soil's cohesive strength, thereby increasing the stand-up time of the excavated face. This chapter discusses the applications of chemical grouting in underground construction, followed by sections on the types of chemical grout, the equipment used to batch and inject chemical grout, the design of the chemical grouting program, and two case histories, one involving grouting from the surface and the other involving grouting from the tunnel working face.

10.1 APPLICATIONS

The majority of underground chemical grouting is performed to aid soft-ground tunneling by modifying the soil properties of the ground surrounding the tunnel. Open-face shield-type tunneling machines, as shown in Fig. 10-1, are commonly used to drive tunnels through soft ground when the tunnel is located above the water table. When cohesionless or "sugar" sand is encountered, problems with running ground can develop at the working face. The running ground condition is serious because, in addition to causing added cost and schedule delays to the tunneling operation, it creates voids in the ground above the tunnel that can lead to surface subsidence and chimneys or sinkholes to the ground surface. Chemical grouting is used to modify cohesionless soil by increasing its cohesive strength, thereby limiting or eliminating the running ground problem. In such cases, chemical grout is injected into the soil, either from the surface or from the tunnel working face, in order to create an arch or full-face of modified soil through which the tunnel can be excavated.

[a]This chapter is contributed by Daniel F. McMaster and Michael J. Robison.

FIG. 10-1. Soft-Ground Tunnel Shield

10.2 TYPES OF CHEMICAL GROUT

Chemical grouting of soil was first used in Europe in the late 1800s, when concentrated sodium silicate was injected into one hole and a coagulant into a second, nearby hole. This process was used with and without modification to grout soils until the early 1950s (Karol 1983). Since then, hundreds of grout formulations have been developed. Many different types of chemical grout, with very specific uses, are available on the market today. These grouts can be divided into the following categories (Karol 1983):

- Sodium Silicate Formulations
- Acrylamides
- Lignosulfonates
- Phenoplasts
- Aminoplasts
- Water-reactive Materials
- Other Chemical Grouts

The underground construction industry in the United States has used sodium silicate formulations almost exclusively to stabilize dry, cohesionless sands for soft-ground tunneling. Sodium silicate–based grouts have been found to perform satisfactorily and to be less expensive than other types of chemical grout in high-production operations. The remainder of this section discusses the properties of sodium silicate grout. Although information on other types of chemical grout is not presented in detail, it can readily be found in the references.

Grouting with sodium silicate involves mixing a dilute solution of sodium silicate with an acid or acid salt reactant to form a gel. The time required for formation of the gel, which is controllable, is a function of the chemical concentrations of the sodium silicate and the reactant. The application of sodium silicate grouting requires combining the gel base with the reactant, with or without the use of an accelerator, just prior to the grout's injection. The gel base usually consists of a 30–40% solution of Grade 40 sodium silicate with a specific gravity of 41.5° Baume, diluted with water.

Reactants usually consist of solutions of formamide, diacetate, or sodium bicarbonate. When used, accelerators such as calcium chloride are added just prior to injection. The process produces a gel with a set time varying from 2–60 min. It should be noted that set time is temperature dependent, therefore, when working in lower-temperature conditions, increasing the concentration of the reactant may be necessary. Sodium silicate–based grouts can develop unconfined compressive strengths on the order of 70–3,500 kPa (10–500 psi), depending primarily on silicate content and set time, but also on reactant, grain size, and other factors.

10.3 EQUIPMENT

The drilling equipment used for chemical grouting is basically the same as that used in cementitious grouting (see section 4.1). The specialized equipment used in the chemical grouting process is the batching and injection equipment.

10.3.1 Batching Equipment

Sodium silicate grout is batched using a two-tank, two-pump delivery system. By using this dual delivery system, the two components that react to make the silicate gel, the sodium silicate and the reactant, are not combined until they reach the grout pipe. This allows for better control of gel time while preventing the grout from prematurely gelling in the mixing tank, the pump, or the delivery lines. The delivery system is illustrated in Fig. 10-2. It is common to have the sodium silicate and the reactant solutions stored in bulk, either in large on-site tanks such as Baker tanks, or in over-the-road tanker trucks parked at the site. From these tanks, the silicate solution and the reactant are metered out and pumped to the grout hole. Metering systems and pressure gauges are installed in each delivery line to monitor and control the flow of the components. Pumping is typically accomplished using progressing helical cavity (Moyno) pumps, however, piston-type pumps are also used. Progressing helical cavity pumps produce a continuous, uniform flow

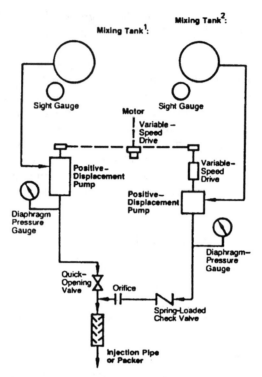

FIG. 10-2. Chemical Grout Delivery System (Karol 1983)

of materials; therefore, better control of the grouting can be achieved. Section 4.5 discusses progressing helical cavity pumps in more detail.

10.3.2 Injection Equipment

The injection equipment consists of grout pipes installed in drill holes, drilled from either the surface or from the tunnel working face. A more sophisticated and widely used chemical grout piping system is the manchette or sleeve-port pipe, which gives the grout injection operator increased control over the specific depth at which grout enters the soil strata. This system was developed in Europe, where it is known as the tube-a-manchette system. The sleeve-port pipe system is illustrated in Fig. 10-3. Sleeve-port pipes are shown in Figs. 2-12 and 2-13. The sleeve-port grout pipe, which is typically 25–50 mm (1–2 in.) diameter and composed of PVC, is inserted into the drill hole. The annular space between the pipe and the wall of the drill hole is filled with a weak grout, usually consisting of cement and bentonite.

CHEMICAL GROUTING 173

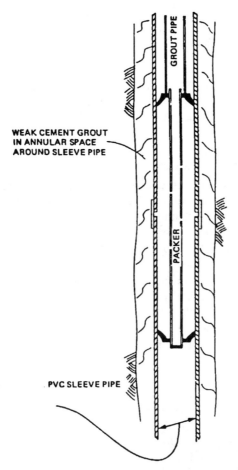

FIG. 10-3. Sleeve-Port Pipe (Karol 1983)

Chemical grout is injected into the pervious soil formation through openings or ports in the sleeve-port pipe. In this process, a double-packer is lowered to the specified depth to isolate the desired port openings. Under pressure from the chemical grout, the rubber tubing, which is located on the outside of the PVC pipe surrounding the injection port, expands. This allows the chemical grout to fracture the weak annular grout between the PVC pipe and the soil. The chemical grout enters the soil through the fractured annular grout. The process is repeated until grouting at all the desired sleeve-port locations is completed. The sleeve-ports can be injected in any sequence or may be reinjected with grout as required.

10.4 DESIGNING THE CHEMICAL GROUTING PROGRAM

Chemical grouting to stabilize cohesionless ground can be applied from the surface or from the tunnel working face. With either method, the goal of a chemical grouting program is to create an arch or face of stable soil through which excavation can proceed. The principal factor affecting the design of a chemical grouting program is ground characteristics. Therefore, geotechnical data, such as grain size distribution, permeability, porosity, and groundwater, must be assessed.

10.4.1 Groutability of Ground

A "groutable" soil is one that will, under practical pressure limitations, accept injection of a given chemical grout at a sufficient rate to make the project economically feasible (Baker 1982). The first step in designing a chemical grouting program is to determine if the ground can be chemically grouted. Grain-size distribution curves can be used to assess the groutability of a granular soil as shown in Fig. 10-4. Soils with less than 10% fines are groutable. Soils with 10–20% fines are moderately groutable. Soils with 20–25% fines are marginally groutable. Soils with more than 25% fines generally are not groutable.

The groutability of the ground can also be assessed on the basis of permeability. Chemical grouting can be performed easily in pervious soils

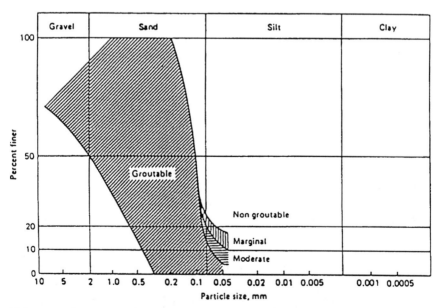

FIG. 10-4. Groutability Based on Grain-Size Distribution (Karol 1983)

having a coefficient of permeability k in the range of 10^{-1} to 10^{-3} cm/s. Soils with permeability in the range of 10^{-3} to 10^{-5} cm/s are moderately groutable, while soils with k values of 10^{-5} to 10^{-6} cm/s are considered impractical to grout. Soils with a permeability of less than 10^{-6} cm/s are generally ungroutable (Karol 1983). Section 3.6, Groutability Ratio, gives a more in-depth discussion of permeability constants.

10.4.2 Grouting Program Design

The second step in designing a chemical grout program, after it has been determined that the ground is chemically groutable, is to design the chemical grout program itself. This involves selecting the type of chemical grout, its strength and gel time, the location of the grout holes and the injection ports, and the volume of grout to be injected at each port.

An important parameter in determining the volume of chemical grout to be injected is the porosity of the soil. Typical groutable soils have porosities ranging from 25–50%. The actual porosity is governed by the grain size distribution and soil density. Porosities for various soils are shown in Table 10-1.

In designing the grouting program, the desired geometry of the zone to be grouted is first established. Next, an injection pipe pattern is developed within the geometry of the zone to be grouted. It is generally assumed that flow from grout injection ports will be radial and uniform. Grout pipes are spaced in a pattern that provides for primary, secondary, and sometimes

TABLE 10-1. Porosities and Grain Size Distribution for Various Soils (NAVFAC 1982)

Soil description (1)	Particle Size and Gradation				Porosity (%)	
	D_{max} (mm) (2)	D_{min} (mm) (3)	D_{10} (mm) (4)	Uniform coefficient CU (5)	D_{max} loose (6)	D_{min} dense (7)
Standard Ottawa sand	0.84	0.59	0.67	1.1	44	33
Clean, uniform sand (fine or medium)	—	—	—	1.2–2.0	50	29
Silty sand	2.0	0.005	0.02	5–10	47	23
Clean, well-graded sand (fine to course)	2.0	0.05	0.09	4–6	49	17
Micaceous sand	—	—	—	—	55	29
Silty sand and gravel	100	0.005	0.02	15–300	46	12
Sandy or Silty clay	2.0	0.001	0.003	10–30	64	20
Gravel, sand, silt, and clay (mixture)	250	0.001	0.002	25–1,000	41	11

tertiary injection. Often the grout pipe locations are split-spaced from one another. Spacing of grout pipes generally ranges from 0.5 to 2.5 m (1.5 to 8 ft). The spacing of injection ports in the grout pipes generally ranges from 0.3 to 1.0 m (1 to 3 ft).

Grout volumes can be calculated once the hole pattern is established. The volume of chemical grout required is calculated using the following equation:

$$V_g = V_z (nF)(1 + L) \qquad (10.1)$$

where V_g = liquid volume of grout; V_z = total volume of treatment zone; n = soil porosity; F = void filling factor; and L = grout loss factor beyond the boundary.

The void-filling factor generally ranges from 0.85 to 1.0 and is generally governed by pore size and percent of fines. The grout-loss factor generally ranges from 0.05 to 0.15 and is governed by the grout zone geometry, number of injection points, and variability in ground conditions.

10.5 CHEMICAL GROUTING CASE HISTORIES

Chemical grouting can be performed from the ground surface or from the tunnel working face. Grouting from the surface in advance of the tunneling operation has the advantage of not interfering with the actual tunneling process. In addition, chemical grouting from the surface can be performed early in the project, sometimes months or even a year in advance, before actual tunneling begins, allowing greater scheduling flexibility. When the depth of the soil to be grouted exceeds approximately 15 m (50 ft), grouting from the surface becomes impractical.

Chemical grouting performed from the tunnel working face has several advantages over chemical grouting performed from the surface. The advantages of grouting from the working face are that it eliminates surface disruptions to commercial and private properties, the need to reroute vehicle traffic, and the cost of relocating utilities. Additionally, when grouting is performed from the working face it can be used on an as-needed basis, which is dictated by the actual soil conditions encountered. By having specific geologic knowledge of the in situ soil conditions, the extent of the area to be treated can be kept to a minimum. The disadvantage of grouting from the working face is that it interferes with the tunneling operations, thus impacting tunneling costs and schedule.

10.5.1 Chemical Grouting from Surface

The MUNI Metro Turnback project in San Francisco is an example of a project where chemical grouting was performed from the surface. The project consisted of 260 m (850 ft) of twin-bore, 5.64 m (18.5 ft) excavated-diameter subway tunnels extending along 97 and 107 m (318 and 350 ft) radius curves. The project provides an extension of the MUNI Metro tunnels from the midlevel of the Embarcadero Station beneath Market Street up to the Embarcadero. The tunnels on this project were mined under compressed air and lined with gasketed, bolted steel liner plates. The depth to tunnel invert ranged from 12.1 to 13.7 m (40 to 45 ft) below the surface.

Soil conditions along the tunnel alignment consisted of Bay Mud overlain by Dune Sand fill. The Dune Sand fill consisted of loose to medium dense, clean fine sand. The Bay Mud consisted of a soft to medium stiff, silty clay with trace amounts of organics, shells, and fine sands. The water table was located approximately 2.4 m (8 ft) below ground surface. The location of the Bay Mud/Dune Sand fill interface relative to the tunnels varied from above the tunnel arch to 0.6 m (2 ft) above the tunnel invert.

The chemical grout program was designed to achieve soil stabilization for both permanent conditions and temporary construction. The program was intended primarily to improve granular soil shear strengths to prevent liquefaction from producing uplift loads on the tunnel liner during earthquakes. With regard to temporary construction, the chemical grouting would help to stabilize the soil at and around the tunnel heading to minimize the loss of compressed air and help to prevent cave-in of the tunnel face in the event of an interruption of compressed air supply during tunneling work. Based on the project geotechnical reports, two zones were identified where the Bay Mud/Dune Sand fill interface dropped below the proposed tunnel spring line. These zones were selected to receive the chemical grout treatment.

A chemical grouting test section of approximately 15 m by 15 m (50 ft by 50 ft) was located within the project's cut-and-cover section, which served as the tunnel shaft. Two different proprietary chemical grout mix designs were tested in 1.8, 2.1, and 2.4 m (6, 7, and 8 ft) grout hole spacings within the test section. Based on the observations made after excavating the test section, the 2.1 m (7 ft) spacing and a mix design were selected for the production grouting areas.

The two production chemical grout zones extended along approximately 107 m (350 ft) of the alignment and were 21 m (70 ft) wide, as illustrated in Fig. 10-5. In section, the grout zones extended from the top of the Bay Mud to approximately 3 m (10 ft) above the proposed tunnel arch, as illustrated in Fig. 10-6.

The grouting program was performed in advance of the mining operations from the street surface. Sleeve-port grout pipes were installed in 152 mm (6 in.) diameter boreholes, and the annular space was filled with a cement-bentonite grout. Grouting patterns were governed by traffic lane close-down restrictions, but generally progressed on alternating grout holes.

Grouting was performed in either two or three stages. The primary stage

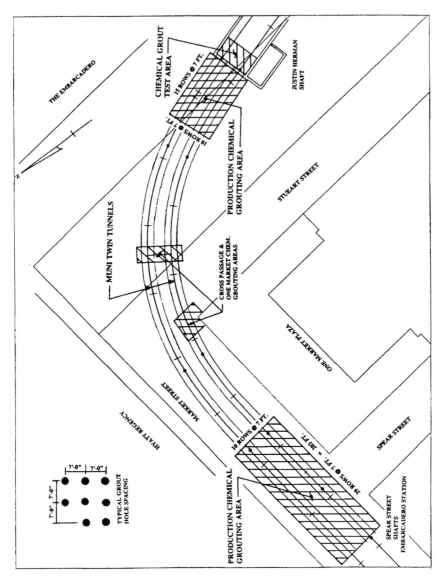

FIG. 10-5. Plan View of Chemical Grout Zone for Muni Metro Turnback

CHEMICAL GROUTING

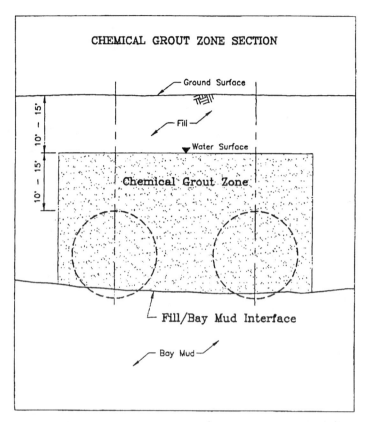

FIG. 10-6. Section of Chemical Grout Zone for Muni Metro Turnback

began at the bottom of each grout pipe and progressed upward at alternating ports, every 1.2 m (4 ft). The secondary stage began 0.6 m (2 ft) above the first primary stage and progressed upward at alternating ports, every 1.2 m (4 ft). A third stage began at the same location as the primary stage and injected grout until a minimum of 105% of the design grout for the 1.2 m (4 ft) interval was met or a high increase in pressure was observed. If the target volume of 105% of the design was accomplished with stages one and two, then the third stage was omitted. The grout injection sequences are illustrated in Fig. 10-7.

10.5.2 Chemical Grouting from Tunnel Working Face

Los Angeles Metro Rail Contract A-146 is an example of a project where chemical grouting was performed from the tunnel working face (Robison et al. 1991). The project consisted of 640 m (2,100 ft) of twin-bore, 6.7 m (22 ft)

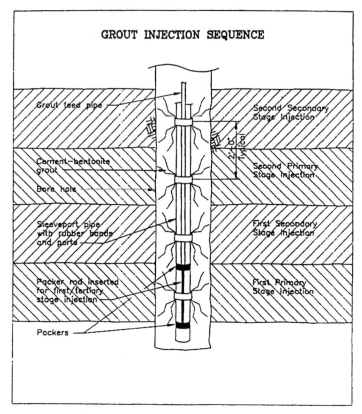

FIG. 10-7. Grout Injection Sequence Used on Muni Metro Turnback

excavated diameter, 5.5 m (18 ft) finished diameter subway tunnels extending along a 300 m (1,000 ft) radius curve between the 5th/Hill Station (Pershing Square Station) and the 7th/Flower Station (Metro Center Station). The tunnels were mined using open-face digger shields that included breasting jacks and a breasting table. The depth to invert ranged from 18.1 to 24.2 m (60 to 80 ft) below the surface.

Soil conditions along the tunnel alignment consisted of firm to very stiff silts and clays and medium dense to very dense clean sands and gravel belonging to the Los Angeles River alluvial deposit. Tunneling began with a firm silt layer in the crown. After mining approximately 250 m (820 ft), cohesionless granular material (sugar sand) was encountered in the crown of the tunnel and a large ground loss occurred. Because a large ground loss could have catastrophic consequences on upcoming buildings the cohesionless sand needed to be stabilized. Due to the lack of surface access and to

CHEMICAL GROUTING

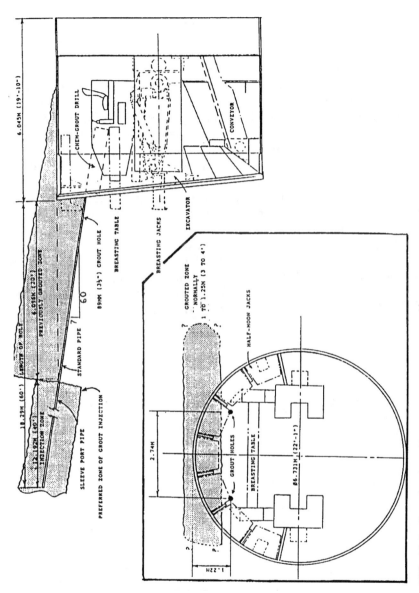

FIG. 10-8. Grouting from Tunnel Working Face on Los Angeles Metro

prevent disruption to overlying facilities, chemical grouting from the tunnel working face was the method chosen.

The chemical grout program was designed to achieve soil stabilization from the tunnel working face, since grouting from the surface would have required access into the buildings along the alignment. The program was intended to allow grouting to occur on the swing and graveyard shifts and for tunnel excavation to occur through the grouted zone on day shifts. The program was designed such that only the ground necessary for face control would be modified by chemical grouting, and, therefore, it was decided to create an arch or canopy over the portion of the tunnel to be excavated.

In performing the grouting, two holes were drilled ahead of the face from the breasting table on the tunnel shield, and sleeve-port pipes were inserted into the holes. Grouting was performed in stages beginning at the far end of the sleeve-port pipe at a depth of 18.3 m (60 ft) and working back to the leading edge of the previous grout zone, which was located at least 6 m (20 ft) ahead of the tunnel face. The working face grouting method used on this project is illustrated in Fig. 10-8. The grout consisted of a solution of 40% Grade 40 sodium silicate mixed at the entry end of the sleeve-port pipe with an organic reactant, either Glyoxal and Diacetin.

REFERENCES

American Concrete Institute. (1967). "Cement and concrete terminology." *Rep., SP-19 (85)*, Detroit, Mich.
American Concrete Institute. (1986). "Guide for cast-in-place low-density concrete." *Rep., ACI 523.1R-86*, ACI Committee 523, Detroit, Mich.
Asphalt Institute. (1978). "Soils manual for the design of asphalt pavement structures." *Manual Series Number 10 (MS-10)*, College Park, Md., ii, 5.
Baker, W.H. (1982). "Planning and performing structural chemical grouting." *Proceedings, Conference on Grouting in Geotechnical Engineering*, American Society of Civil Engineers, New York, N.Y.
Baker, W.H. (1985). "Embankment foundation densification by compaction grouting." *Proceedings, Issues in Dam Grouting*, W.H. Baker, ed., American Society of Civil Engineers, New York, N.Y., 104–122.
Baker, W.H., Cording, E.J., and MacPherson, H.H. (1983). "Compaction grouting to control ground movements during tunneling." *Underground space*, Vol. 7, Pergamon Press Ltd., Oxford, U.K., 205–212.
Brady, G.S., and Clauser, H.R. (1986). *Materials handbook*. McGraw-Hill, New York, N.Y.
Bruce, D.A. (1995). Personal communication.
Bruce, D.A., Granata, R., Mauro, M., and Cippo, A.P. (1993). "Some recent developments in ground treatment for tunnelling." *Proceedings, Rapid Excavation and Tunneling Conference*, Society for Mining, Metallurgy, and Exploration, Inc., Cushing-Malloy, Inc., Ann Arbor, Mich.
Burke, J. (1995). "Success at Islais Creek." *World tunneling and subsurface excavation*. The Mining Journal, Ltd., London, U.K.
Cambefort, H. (1977). "The pricipals and applications of grouting." *Quarterly Journal of Engineering Geology*, 10(2), 57–95.
Clarke, W.J. (1995). Personal correspondence.
Continental Drilling Company. (1993). Personal correspondence.
Dienhard, H., Prinz, H., and Zeidler, K. (1991). "Variations on an NATM theme." *Tunnels and Tunneling*, 23(11), 49–51.
Doig, P.J. (1985). "Grouting and freezing for water control in Milwaukee." *Proceedings, Rapid Excavation and Tunneling Conference*, Vol. 2, American Institute of Mining, Metallurgical, and Petroleum Engineers, Inc., Port City Press, Baltimore, Md., 1211–1224.
Dr. G. Sauers Corporation. (1995). Personal correspondence.
Droof, E.R., Tavares, P.D., and Forbes, J. (1995). "Soil fracture grouting to remediate settlement due to soft ground tunneling." *Proceedings, Rapid Excavation and Tunneling Conference*, Society for Mining, Metallurgy, and Exploration, Inc., Cushing-Malloy, Inc., Ann Arbor, Mich., 21–40.

Ewart, F.K. (1958). *Rock grouting with emphasis on dam sites.* Springer-Verlag, Berlin, Germany.

Glossop, R. (1961). "The invention and development of injection processes. Part II: 1850–1960." *Géotechnique,* London, England, Vol. 11, 255–279.

Goff, J.S., Navin, S.J., and Moulton, W.W. (1985). "Grouting of the bird's nest aquifer at the river shale project." *Proceedings, Rapid Excavation and Tunneling Conference,* Vol. 2, Society for Mining, Metallurgy, and Exploration, Inc., Port City Publishers, Baltimore, Md., 1171–1190.

Gonano, L.P., and Sharp, J.C. (1984). "Design and prestress grouting of a concrete lined high pressure tunnel at Drakensburg." *Innovative Cement Grouting, Publication SP-83-2,* American Concrete Institute (ACI), Detroit, Mich., 19–41.

Goodman, R.D., Moye, S.A., and Javandel, I. (1965). "Groundwater Inflow during Tunnel Driving." *Engineering geology,* Vol. 2, 39–56.

Heuer, R.E. (1995). "Estimating rock tunnel water inflow." *Proceedings, Rapid Excavation and Tunneling Conference,* Society for Mining, Metallurgy, and Exploration, Inc., Cushing-Malloy, Inc., Ann Arbor, Mich., 41–60.

Hoek, E., and Bray, J.W. (1977). *Rock Slope Engineering.* Institute of Minerology and Metalurgy, London, U.K.

Houlsby, A.C. (1982). "Optimum water:cement ratios for rock grouting." *Proceedings, Conference on Grouting in Geotechnical Engineering,* American Society of Civil Engineers, New York, N.Y., 317–331.

Houlsby, A.C. (1985). "Cement grouting: Water minimizing particles." *Proceedings, Issues in Dam Grouting,* W.H. Baker, ed., American Society of Civil Engineers, New York, N.Y., 34–75.

Houlsby, A.C. (1986). "Cement grouting: Reference material." *Eight Annual Short Course on Fundamentals of Grouting,* University of Missouri, Rolla, Mo.

Hunt, R.E. (1984). *Geotechnical engineering investigation manual.* McGraw-Hill, New York, N.Y.

Jumikis, A.R. (1983). *Rock mechanics,* 2nd Ed., Trans Tech Publications, Germany, 37.

Karol, R.H. (1983). *Chemical grouting.* Marcel Dekker, Inc., New York, N.Y.

Klein, A., and Polivka, M. (1958). "Cement and clay grouting of foundations: The use of admixtures in cement grouts." *Journal of Soil Mechanics and Foundation Engineering Division,* American Society of Civil Engineering, 84(1), 1547-1–1547-24.

Lombardi, G. (1985). "The role of cohesion in cement grouting of rock." *Proceedings, Fifteenth Congress on Large Dams,* 3(Q58, R 13), 235–261.

Lyman, T.J., Robison, M.J., and Lance, D.S. (1988). "Compaction and chemical grouting for ground stabilization for the Papago Freeway drain tunnels in Phoenix, AZ," *Proceedings, Second International Congress of Case Histories in Geotechnical Engineering,* St. Louis, Mo.

MacGillivray, D.M. (1979). "High speed shaft sinking." *Proceedings, Rapid Excavation and Tunneling Conference*, Vol. 2, Society for Mining, Metallurgy, and Exploration, Inc., Port City Publishers, Baltimore, Md., 1197–1216.

Mearl Corporation. (1992a). *Technical Bulletin G-103*, Roselle, N.J.

Mearl Corporation. (1992b). *Publication GM600 4.92*, Roselle, N.J.

Mitchell, J.K. (1970). "In-place treatment of foundation soils." *Journal of the Soil Mechanics and Foundations Division*, American Society of Civil Engineers, 96(1), 73–110.

Moller, D.W., Minch, H.L., and Welsh, J.P. (1984). "Ultrafine cement pressure grouting to control ground water in fractured granite rock." *SP 83-8*, American Concrete Institute, Detroit, Mich., 129–151.

NAVFAC (1971). "Soil mechanics, foundations and earth structures." *Design Manual DM-7*, Naval Facilities Engineering Command, Alexandria, Va.

Nel, P.J.L. (1981). "Shaft sinking concepts and two independent systems of concurrent development through shafts being sunk." *Proceedings, Rapid Excavation and Tunneling Conference*, Vol. 2, Society for Mining, Metallurgy, and Exploration, Inc., Port City Publishers, Baltimore, Md., 976–995.

O'Rourke, T.D., (ed.). (1984). *Guidlines for tunnel lining design*. Technical Committee on Tunnel Lining Design, Underground Technology Research Council, American Society of Civil Engineers, New York, N.Y.

Peck, R.B. (1969). "Deep exacavations and tunnelling in soft ground." *Proceedings, Seventh International Conference on Soil Mechanics and Foundation Engineering*, State-of-the-Art Volume, Mexico City, Mexico.

Peck, R.B., Hanson, W.E., and Thornburn, T.H. (1974). *Foundation Engineering*, 2nd ed., John Wiley and Sons, New York, N.Y.

Proctor, R.V., and White, T.L. (1977). *Earth tunneling with steel supports*. Commercial Shearing, Inc., Youngstown, Ohio.

Robinson, M.J., and Wardwell, S.R. (1991). "Chemical grouting to control ground losses and settlements on Los Angeles Metro Rail Contract A146." *Proceedings, Rapid Excavation and Tunneling Conference*, Society for Mining, Metallurgy, and Exploration, Inc., Cushing-Malloy, Inc., Ann Arbor, Mich., 179–195.

Rosenbaum, D. (1991). "Concrete was never so complicated." *ENR, Jan. 21*, McGraw-Hill, New York, N.Y., 35–40.

Smith, D.K. (1987). "Cementing." *Monograph*, Vol. 4, Society of Petroleum Engineers, New York, N.Y.

Terzaghi, K., and Peck, R.B. (1967). *Soil mechanics in engineering practice*. 2nd Ed., John Wiley and Sons, New York, N.Y.

Tirolo, V. (1994). "Shield and compressed air tunneling in the nineteenth century." *One hundred twenty years of tunneling in New York City*, American Society of Civil Engineering, New York, N.Y.

U.S. Army Corps of Engineers. (1984). *Engineering Manual EM 1110-2-3506*, Washington, D.C.

Vaughan, P.R. (1963). "Discussion." *Grouts and drilling muds in engineering practice*, R. Glossop, ed., Butterworth, Inc. Washington, D.C.

Water Resources Commission of New South Wales (1981). "Grouting manual." *Publication 624.159.4G/WAT/16*, Water Resources Commission of New South Wales, North Sydney, Australia.

Weaver, K.D. (1991). *Dam foundation grouting*. American Society of Civil Engineers, New York, N.Y.

Welsh, J.P. (1991). "Chapter 8: Grouting." *Underground Structures*, R.S. Singha, ed., Developments in Geotechnical Engineering series, Vol. 59, Elsevier, New York, N.Y.

Weyerman, W.J. (1958). "Rockfill Dams; the Paradela Dam—foundation treatment." *Journal of the Power Division*, American Society of Civil Engineers, 84(4), 1748-1–1748-9.

INDEX

Accelerators 126–127
Acrylamides 170
Admixtures 124–127, 149
Agitators 96, 98
Alluvial 15, 180
Ascending stage grouting
 (see Stage grouting)
ASTM 121, 125, 126, 127,
 130–131, 164, 167
Automated batching systems
 113–116
Backer tank 171
Backfilling 11, 43, 49
Backfill grouting (see Contact
 grouting)
Backpack grouting (see Contact
 grouting)
Baltimore Regional Rapid Transit
 System 12
Batching equipment, chemical
 grouts 171–172
Bay mud 8, 177
Bedding planes 23, 67, 140
Bentonite 11, 17, 20, 128, 148,
 149, 172
Bits (see Drill bits)
Blast furnace slag 123
Blaine fineness 123
Blocky rock 2
Bolton Hill Tunnel 12
Bore hole logs 70, 72, 73
Boston Commonwealth Pier 51
Boston Harbor Inter-Island Tunnel
 61, 64
Breasting shelves (boards) 15, 16,
 182
Calcium chloride 126
Cast-in-place concrete lining 5,
 23, 24, 26–30, 36, 39, 44,
 46, 52, 108

Cast iron lining 1, 24, 37
Cavities 23, 123
Cellular concrete 11, 34, 36, 49–52
Cement, Portland 11, 17, 43,
 121–122, 124; gradation
 curve 17, 19
Cement, Ultrafine 17, 43,
 123–124; gradation curve
 17, 19
Certification 162
Chambers 1, 24, 32, 60, 64, 87
Chemical additives (see
 Admixtures)
Chemical grouting 15, 17,
 169–182
Chimneys 15, 16
Clay 8, 12, 67, 177
Climatic factors (temperature) 30,
 126, 171
Cobble 15
Colloidal mixers (see Mixers)
Compaction grouting 6, 11–16,
 65; hole spacing 11, 15, 16;
 mixes 11, 16; pressure 11, 15
Compatibility 167
Compressed air tunneling 10, 11,
 15, 76, 177
Concrete batch plant 98
Consolidation grouting 23, 30,
 32, 53, 65, 84, 86, 105, 133,
 140–142, 155
Contact grouting 5, 6, 24–30, 32,
 34, 36, 39, 43, 65, 84, 86,
 93, 105, 133–140, 148, 155
Core holes 90
Cost 1, 2, 52, 65, 176
Cracks 30
Crew sizes 155–157
Curtain grouting 6, 23–24, 65,
 84, 133, 142–145, 155

Data acquisition and recording equipment 110–113, 171–172
Delivery and distribution system 109–110
Density test 150, 167
Displacement 12
Dispersant 126
Disturbed soil 11
Down-hole hammers (see Drilling equipment)
Drill-and-blast 32, 52, 60, 86, 140
Drill bits 86, 90, 92
Drilling equipment 84–93; percussion 85, 86–90; rotary 84, 90–91; down-hole-hammer 84, 92–93
Drilling logs (see Drilling reports)
Drilling reports 164–165
Drop-pipes 98
Earth pressure balance (EPB) machine 22, 35, 76
Embedment grouting 6, 43–44, 65
Equipment 2, 6, 83–120, 171–173
Equipment configuration 117–120, 171–172
Existing structures 11, 12, 16, 22, 37
Extensometers 48
Fast ravelling ground 76
Faults 23, 67, 140
Feeler holes (see Probe holes)
Field testing 167
Fillers (see Sand)
Flowing ground 76
Fluidifiers (see Dispersants)
Fly ash 11, 16, 17
Fracture aperture 81
Fractures 67
Gauge savers 103–104
Gas-producing agents 127
Geotechnical investigation 23, 66–70, 129

Geotechnical design summary report (GDSR) 66–70
Grain size (see cement)
Gravel 11, 12, 15, 34, 68
Ground modification 16, 61, 169
Ground runs 15, 16
Groundwater 1, 2, 6, 15, 22, 23, 30, 37, 52, 61, 65, 70–75, 122, 174
Grout agitators (see Agitators)
Grout curtain (see Curtain grouting)
Grout durability (leach) 148
Grout headers 109
Grout hole layout 133–145
Grout holes, deviation (see Hole deviation)
Grout holes, diameter and depth 2, 145
Grout holes, orientation 2, 133
Grout holes, spacing 2, 133, 134, 140, 142
Grout leaks 32
Grout mixes 11, 17, 30, 80, 102, 126, 182
Grout mixers (see Mixers)
Grout mixing 100
Grout plants 96
Grout placement operations 147–160
Grout pipes 40, 41 (also see Nipples)
Grout pumps (see Pumps)
Grout rings 133, 140, 142, 143
Grout take criteria (see Refusal criteria)
Grout testing, field (see Field testing)
Grout testing, laboratory (see Laboratory testing)
Grout waste 2, 169–170
Groutability 17, 81, 174

Groutability ratio 17, 80–81, 174–175
Grouting ahead of excavations 61–64
Grouting in rock 22–24
Grouting in soil 6–22
Grouting jumbos 117
Grouting materials 121–128
Grouting pressure (see Injection pressure)
Grouting procedures (see Grout placement operations)
Grouting to increase stability 66, 75–80
Grouting to increase strength 23, 66, 80
Grouting to limit groundwater infiltrations 70–75
Hallandsasen Railway Tunnel 61, 64
Hardrock 6, 34
Headers (see Grout headers)
Helms Pumped Storage Project 162
History 1
Hole deviation (hole wandering, hole alignment) 85, 86, 88
Hoover Dam 1
Hydraulic routing 81–82
Hydroelectric 2, 6, 38, 83
Hydrofracture grouting 6, 21–22, 65; hole spacing 22; pressure 21, 22
Hydrogen sulfide 37
Injection pressure 2, 30, 37, 102, 103, 110, 149–155
In situ pressure testing 67
Inspection 30, 162–164
Instrumentation 22
Islais Creek Project 8
Jet grouting 6–11, 65
Joints 23, 53, 67, 70, 123, 124, 140, 148

Laboratory testing 67, 164
Loads from grouting 12, 149–155
Los Angeles Metro Rail 179
Lugeon test 68
Loose soil 2, 11, 177
Manchette tube 20, 21, 172, 182
Mapping cracks in concrete linings 30–32
Marsh viscosity test 167
MC-100 123–124
MC-300 123–124
MC-500 123–124
Membrane liners (see Waterproofing system)
Methane gas 37
Methods specifications 2, 129
Microfine cement (see Ultrafine cement)
Mixers 93–98; paddle 93, 148; colloidal 93–148
Mixed-face tunneling 6
Mobile platforms (see Grout jumbos)
Monitoring (soil) 15, 22
Moran car 98, 116
Moyno pump (see Pumps)
Multiple row grout curtains 142
Muni Metro Turnback 176
New Austrian Tunnel Method (NATM) 19, 39
New Croton Dam 1
Nipples 51, 107–108
North Outfall Replacement Sewer 51
Open-faced shield 15, 35, 36, 169, 180
Packers 20, 57, 68, 104–107, 145, 150, 172
Paddle mixers (see Mixers)
Papago Freeway Drainage Tunnel 15
Pea gravel 34, 36

Penstock (see Steel penstock lining)
Percussion drills (see Drilling equipment)
Performance specification 2, 11, 16, 129
Permeability 66, 67, 70, 73, 81, 82, 155, 174, 175
Permeability tests 67, 68, 110
Permeation grouting 6, 17–21, 65; hole spacing 20; mix 17; pressure 21
Piston pump (see Pumps)
Poling plates 16
Porosity 81, 82, 174, 175
Portal 5, 8, 19, 29, 99, 113, 145, 154
Powerhouses 24, 28, 83, 145
Precast concrete lining 24, 32–36, 102
Pressure gauges 103
Pressure tesing 67
Prestressing 6, 44–49, 65
Pricing 23, 30
Primary lining 32, 35, 36, 37, 39
Primary support 10, 11, 34, 35, 37
Probing and grouting ahead of tunnel and chamber excavation 60–64, 84
Probe holes 60–61
Production 155–157
Progressing helical pump (see Pumps)
Proportioning 147–149
Pumps 96, 101–103, 171
Quality (general) 2, 83
Quality (control) 161–168
Recording equipment (see Data acquisition and recording equipment)
Records (see Inspection)
Refusal criteria 2, 30, 155
Reservoirs 23, 24, 145

Rings (see Grouting and segments)
Roadheader 11, 23, 39, 60
Rock 6, 22, 23, 24, 32, 39, 67, 70, 81, 123
Rock (definition) 5
Roughness 81, 82
Running soil (running ground) 2, 76, 169
Saint Clair River Tunnel 22
Safety and environmental issues 157–160
Sand 8, 11, 12, 15, 67, 170, 177, 180
Sanded grouts 16, 17, 102, 124, 148–149
Schedule 1, 2, 34, 57, 64, 83, 91, 162, 176
Schurzeburg Tunnel 17
Secondary liner 32
Segments 32, 33, 34, 36
Set time 126
Settlement (ground) 15, 21, 22
Settlement (grout mix) 121, 167
Shaft collar 53–56
Shaft grouting 23, 52–59, 84; pregrouting from the surface 53–56; in-shaft grouting 56–59
Shafts 1, 22, 23, 32, 36, 38, 44, 83, 113, 154
Shears 67
Shield 1, 10, 11, 15, 16, 35, 37, 60, 169, 182
Silica gel (see Sodium silicate)
Silt 8, 11, 12, 67, 177
Slag (see Blast furnace slag)
Sleeve tubes (see Manchette tube)
Slow ravelling ground 76
Slurry shield 35, 76
Sodium chloride 126
Sodium hydroxide 126
Sodium silicate 170, 182
Soft ground 6, 36, 39

INDEX

Soft-ground tunnel 8, 9, 15, 22, 32, 34, 36, 37, 76, 169
Soil 6, 8, 11, 15, 17, 21, 24, 35, 65, 76, 169
Soil (definition) 5
Soil fracture (see Hydrofracture grouting)
Soil gradation curves 17, 19
Soil modification 10, 16, 49
Specialty contractors 11, 16
Specific gravity 167
Specifications 2, 10, 11, 15, 16, 22, 23, 30, 43, 60, 86, 121, 124, 129–132, 155, 162
Split-spacing sequence 24, 26, 54, 142
Squeezing ground 10
Stability (ground) 75–80
Stabilizers (see Bentonite)
Stage grouting, ascending (stage-up) 53, 105, 177–179, 182
Stage grouting, descending (stage-down) 53, 105
Steel and cast iron segments 24, 36–37
Steel penstock lining 38–39, 43, 46, 134
Steel pipe 11, 36
Steel ribs and lagging (steel sets and lagging) 11, 24, 28, 36, 37
Structural grouting 24–52

Sulfate resistant grouts 122
Super fine cement (see Ultrafine cement)
Superplasticizers (see Dispersants)
Supervision 155
Surface leakage 53
Test grouting 164–168
Test reports 162
Testing (see Field testing and laboratory testing)
Tortuosity 81
Transit mix trucks 98
Tube-a-manchette (see Manchette tube)
Tunnel Boring Machine (TBM) 12, 23, 32, 33, 39, 60, 76
Ultrafine cement (see Cement)
Valves 109, 112
Vicat needle 167
Viscosity 167
Voids 1, 5, 24, 26, 29, 33, 34, 35, 43, 44, 65, 93, 102, 124, 133, 134, 136, 148
Wash water 2, 159, 160
Water (in grout mix) 11, 16, 17, 124, 127–128
Water:cement ratio 30, 124, 147–148, 150
Water inflows 10, 15
Water meters 100–101
Water reducing agents 11, 126
Waterproofing systems 39–43
Wells (dewatering) 15